G. F. Rodwell

The birth of chemistry

G. F. Rodwell

The birth of chemistry

ISBN/EAN: 9783741172359

Manufactured in Europe, USA, Canada, Australia, Japa

Cover: Foto ©Angelika Wolter / pixelio.de

Manufactured and distributed by brebook publishing software
(www.brebook.com)

G. F. Rodwell

The birth of chemistry

JOHN MAYOW.
BORN 1645.—DIED 1679.
(From his *Tractatus Quinque Medico-Physici*, 1674.)

THE

BIRTH OF CHEMISTRY.

BY

G. F. RODWELL, F.R.A.S., F.C.S.

Science Master in Marlborough College ;
Late Lecturer on Natural Philosophy in Guy's Hospital.

WITH NUMEROUS ILLUSTRATIONS.

London:

MACMILLAN AND CO.

1874.

PREFACE.

I HAVE endeavoured in the following pages to trace the rise and early development of a very old Science, mainly that we may mark the attitude of thought which actuated the scientific mind in bygone times, and may thus be led to compare the ancient with the modern method of evolving ideas, and building them up into a connected whole. With this object in view I have chosen the earlier history of the Science of Chemistry, in its various phases of: (*a*) primitive theories affecting the history of matter; (*b*) metallurgical chemistry of the ancients; (*c*) alchemy; (*d*) early ideas respecting the nature of combustion; and (*e*) the rise of pneumatic chemistry. The survey has been carried no farther than the time of the fathers of modern chemistry, Lavoisier, Priestley, Scheele, Bergman, Black, Cavendish, and Davy. The labours of these men belong to the later history of the Science.

G. F. RODWELL.

MARLBOROUGH,
Nov. 24*th*, 1873.

CONTENTS.

CHAPTER I.

CHAPTER II.

CHAPTER III.

CHAPTER IV.

CHAPTER V.

CHAPTER VI.

CHAPTER VII.

CHAPTER VIII.

CHAPTER IX.

CHAPTER X.

CHAPTER XI.

LIST OF ILLUSTRATIONS.

THE BIRTH OF CHEMISTRY.

" Quod si quis ad intuendum ea, quæ magis curiosa habentur quam sana, animum submiserit, et alchymistarum aut magorum opera penitius introspexerit, is dubitabit forsitan, utrum risu, an lacrymis potius, illa digna sint. Alchymista enim spem alit eternam, atque ubi res non succedit, errores proprios reos substituit; secum accusatorie reputando, se aut artis aut authorum vocabula non satis intellexisse, unde ad traditiones et auriculares susurros animum applicat aut in practicæ suæ scrupulis et momentis aliquid titubatum esse; unde experimenta in infinitum repetit; ac interim quum inter experimentorum sortes, in quædam incidit aut ipsa facie nova, aut utilitate non contemnenda; hujusmodi pignoribus animum pascit, eaque in majus ostendat et celebrat; reliqua spe sustentat. Neque tamen negandum est alchymistas non pauca invenisse, et inventis utilibus homines donasse."

FR. BACON, *" Novum Organum."*

BIRTH OF CHEMISTRY.

CHAPTER I.

Introduction—Ancient Science—Origin of Chemistry—Derivation of the Name—Definitions of Chemical Science.

THE history of a natural science resembles in many respects the history of a nation. In each instance the object is first to obtain a knowledge of causes, then to frame laws. The first are those causes which most promote the well-being of the nation, the second those causes which produce the phenomena of the Universe. In each instance we start with an absence of all law, and we may observe the slow efforts of the human mind to trace each effect to its proper cause, to group together causes, and finally to connect them by one bond. The main difference is this, that in the case of the nation man has to deal with laws which must be founded upon a just study and close observance of every phase of that particular community, influenced as it is by numberless external causes, such as race, climate, religion, habit of thought, tradition ; while in the case of the science he has to

B

evolve pre-existent laws, also by the close observance of facts, which are hidden from him by the complex mechanism of nature. M. Taine would tell us that the laws which influence the development of peoples are just as absolute, definite, and pre-existent, as those which govern the affairs of nature; but we are quite disinclined to admit this, even in regard to one particular race, in one particular locality. In both histories we have similar forms of government, similar assemblies of lawgivers; we have our aristocracies, oligarchies, democracies, republics: we have at some period or other Conservatives and Liberals of every shade. We know not what Conservative rule can compare with the dominance of the science of Aristotle for twenty centuries, and we cannot be too ready to welcome the Liberal-conservative era of Copernicus and Giordano Bruno, the Liberal era commenced by Galileo and Francis Bacon, which by easy stages is passing, if it has not passed, into the right Radical era of modern scientific thought. The "Republic of learning" is no empty phrase.

No one would venture to deny the value of a knowledge of the history of nations, and we are inclined to believe that the history of the natural sciences is not without its uses. It is neglected because during the last century new discoveries have quickly succeeded each other, old sciences have augmented, while new sciences have arisen; in fact, the progress of science has been so extraordinarily rapid that we have scarcely time to turn aside and look at its past history; the present is sufficient for us, and if we once get out of the main current of thought we have difficulty in regaining lost ground. Yet we may no more forget that we owe our present wise laws and great constitutional system to the labours of ten centuries of men, than that our science of to-day represents the accumulation of the scientific thought of twice ten centuries. Intellectual revolutions have not been less frequent than social revolutions, nor battles of the pen than battles of the sword; the crash of a fallen philosophy has often been louder than that of a fallen throne; the wail of the last Phlogistians rent the heavens; the Aristotelian physics died with groanings, and gaspings, and a discoloured visage.

In tracing the history of a science, we are first led to inquire whether the ancients possessed any knowledge of it, and whether it originated among them. Now, the ancients made but little progress in any of the natural sciences. They divided all human knowledge into three parts: Logic, or mental philosophy; Physics, or natural philosophy; Ethics, or moral philosophy. Some placed logic first, some ethics, but no one physics. Philosophy was compared to an egg—logic the shell, physics the white, ethics the yolk; or, again, it was compared to a living creature—logic the bones, physics the flesh, ethics the soul. Plato separates logic as the knowledge of the immutable, from physics the knowledge of the mutable. The Cynics sought a complete freedom from any object or aim in life, and renounced all science. Sokrates aimed at logical definition, and affirmed that the true nature of external objects can be discovered by thought without observation. The knowledge of one's self ($\gamma\nu\hat{\omega}\theta\iota$ $\sigma\epsilon\alpha\nu\tau\acute{o}\nu$) is the true object·

and aim of all philosophy. Others of the ancients allowed that geometry might be employed for the measurement of land, and astronomy cultivated so far as it might be of use to sailors, but on no account as serious subjects of mental occupation. Knowledge obtained from external sources is worthless; there is nothing to be learned from fields and trees. A certain philosopher is said on this principle to have put his eyes out, in order that his mind might not be influenced by external objects, and might be left to pure contemplation. (How curiously this contrasts with the plaint of Galileo just before his death : "*Proh dolor!* the sight of my right eye, that eye whose labours, I dare say it, have had such glorious results, is for ever lost. That of the left, which was and is imperfect, is rendered null by a continual weeping.")

Thus it happened that natural science made but little progress among the ancients; thus it happens that a schoolboy of twelve knows more about earth, and fire, and water, than was dreamt of in the philosophies of the greatest thinkers of antiquity. Let us, however, give them their due; let us confess that Plato possessed the "finest of human intellects, exercising boundless dominion over the finest of human languages;" that Aristotle was the greatest genius the world has ever seen; that as pure intellectual evolutions they have handed down to us a mass of grand philosophy; ten thousand noble efforts of the human spirit. Everything favoured the exercise of the unaided intellect, while it is hard to estimate the difficulties which presented themselves in the investigation of nature. At one period it was

considered impious to attempt to explain the manifestations of the gods.   There was an outcry in
Athens, a popular demonstration, when the thunderbolts of Zeus were referred to common fire produced by the collision of clouds.   The feeling was
of the same nature as that conveyed by Campbell's
stanza :—

> " When Science from Creation's face
> Enchantment's veil withdraws,
> What lovely visions yield their place
> To cold material laws ! "

Only the feeling existed in an intensified form, for
here the first of the gods was derided—the Olympian
Zeus, Lord of the Air, he who rides upon the storm,
and hurls the thunderbolt.   For a length of time,
therefore, any investigation of nature was impossible
for religious reasons.   Men were to worship nature,
to be filled with awe and wonder—δεισιδαιμονία—
in presence of great natural phenomena, but not
to inquire too closely into their causes.   Twenty
centuries later the Doctors of Salamanca who inter-
rogated Columbus, the Inquisitors of the Sacred
College who examined Galileo, upheld the same old
doctrines, albeit the old gods had passed away.   But
the investigation of nature was impossible among the
Greeks ; their capabilities were very limited, they had
no instruments for observations, or experiments of
any kind, neither had they the faculty of observation ; their minds were untutored in that particular
direction.   Then they had to contend against their
own particular habit of thought, the extreme tendency to concretion, to hasty generalisation from

purely mental premises; or if an observation had
been made, a broad general law was deduced from it
without further observation.   So also the Chaldæans
and Parsis had to contend against the mysticism, the
astrology, and magic, which originated among them ;
and the ancient Hindu was so given to extreme
abstraction, and to the evolution of all manner of
strange metaphysical dogmas, that we could scarcely
look for much science from an Eastern source.
Egyptian learning was monopolised by the priests,
and they so wove together the real and the unreal,
and were so secret withal in their actions, that
although much of the Greek learning came direct
from Egypt, we cannot trace it to its direct source,
or point to one Egyptian writer on philosophy.   The
Greeks, too, received much from the Phœnicians ;
but here also we find no record.   We will presently
inquire more fully into the exact amount of science
possessed by the ancients.

    We have chosen for our historical survey one of
the oldest of the natural sciences, for obvious reasons,
the chief one being that it will enable us to observe
more minutely the early thought of ancient peoples
in regard to certain phenomena of nature.   The
science of Chemistry does not owe its existence to
any one people, or to any sudden process of development.   The time when the foundation-stone was
laid is too remote to be even suggested ; the basis
of the edifice is sunk deep in Eastern soil ; the
walls were slowly and laboriously raised during the
Middle Ages, and were completed by Lavoisier,
Black, and Priestley ; the men of our day are work-

ing at the roof. We neither hold with M. Goguet
that Moses possessed considerable knowledge of
chemistry because he dissolved the golden calf, nor
with M. Wurtz, when he says "La chimie est une
science Française. Elle fut instituée par Lavoisier
d'immortelle mémoire." Chemistry was not a science
until long after the time of Moses ; it was a science
long before the time of Lavoisier. We wonder what
Dr. Hermann Boerhaave of Leyden (whose large
quarto "Elementa Chemiæ" was published in 1732,
nine years before the birth of Lavoisier), would say
to the proposition of M. Wurtz. Short of this, it
would be difficult to overrate the services which
Lavoisier rendered to chemistry. But the science
has grown up by a gradual process of evolution ;
upon its surface we find the impress of many and
diverse phases of thought and of action ; the science
of to-day is the summation of many intellectual
efforts produced by the constant struggle of the
human mind for truth. How often that truth has
been hidden by a mass of sophistries ; how often it
has been absorbed by some false philosophy to appear
again untarnished in due time ; how often the attempt
has been made to crush it under foot ; and how it
has ever risen to the surface at last, all who read the
history of faiths, nations, ideas, must know. It will
be our object to show this is the study of the particular science which now engages our attention.

The word χημεία first occurs in the Lexicon of
Suidas, a Greek writer of the eleventh century ; he
defines it as "the preparation of gold and silver."
In the "Lexicon Græco-Latinum" of Robertus Con-

stantinus, published in 1592, the same definition is given, and Suidas is quoted as the authority. According to Olaus Borrichius, however, there were Greek writers on alchemy before this date; there is said to be a Greek MS. of the fifth century on alchemy in the King's Library in Paris, and others of a somewhat later date in the libraries of Munich, Milan, Venice, Hamburg, and Madrid; but we are inclined to doubt whether any of these were written before the ninth or tenth century. They are probably the work of monks living at Alexandria and Constantinople; indeed, one of them is entitled, " Cosma the Monk, his Interpretation of the Art of making Gold." The titles of some of the others will prove to us that we can place but little faith on any date which may be assigned to them :—

"Heliodorus on the Art of making Gold" (περὶ χρυσοποιήσιος).

• "John the High Priest, in the Holy City, concerning the Holy Art."

" Isis the Prophetess to her son Orus."

"Moses the Prophet on Chemical Composition" (περὶ χημευτικῆς συντάξιος).

" Cleopatra on the Art of making Gold."

"Democritus the Abderite, the Natural Philosopher, on the Tincture of Gold and Silver, and on Precious Stones and Purple."

Equally worthless, we believe, are the Greek derivations of the word " chemistry." Many (among others M. Hoefer) derive the word from χέω, to fuse or melt, because the majority of old chemical operations were effected by fire—witness calcination, ignition, distillation, sublimation, desiccation, reverberation. The earliest chemical arts, such as the smelting of metals and the production of glass, were also operations of fire. Indeed, the science has been called *Pyrotechnia* (πῦρ τέχνη, the art of fire), because, says Lemery, in his "Cours de Chimie," "we in effect produce all chemical operations by means of fire." Others derive chemistry from χῦμα—that which is poured out, a liquid, in allusion to the various liquids used in chemical operations; but this derivation is not worth a moment's notice. We must rather look to an Egyptian source. Plutarch tells us that Egypt was called *Chemia*, on account of the black colour of the soil, and that the same term was applied to the black of the eye, which symbolizes that which is obscure and hidden. This word is related to the Coptic *khems* or *chems*, which also signifies obscure, occult, and is connected with the Arabic *chema*, to hide. It is probable that we have here the true derivation of the word chemistry. The first treatise on the science, the date of which is known with any certainty, was written by the Arabian Yeber or Geber, and at that time (the eighth century), Arabic learning had considerable influence on European culture. The science was called the *occult*, or *hidden*, because it related principally to the secret art of the transmutation of metals, as the definition of Suidas, given above, and the earlier works on the science, prove. The term *black art* has been applied both to alchemy and to the magical arts so often associated with it, and clearly agrees with the above derivation. The *al* in alchemy is the Arabic particle *the*, so that alchemy signifies "the hidden science" *par excellence;* we notice the same prefix in *al*koran, *al*cohol (the burning liquid), *al*kali (the acrid substance), *al*gebra, *al*embic (the cup-shaped vessel), and in the names of many stars, as Aldebaran, Algenib, Alpheratz,—all words of Arabic origin.   ·

Whatever difficulties there may be in determining the precise derivation of the word chemistry, there can be none in defining the science as distinctly and definitely the science which treats of the *changes* which matter undergoes; while physics proper treats of the action of various forces—heat, light, electricity, magnetism—upon matter, in all cases unaccompanied by any change of composition. If we heat a piece of iron to redness, or cause it to convey an electric current, or place it in contact with a magnet, it has been submitted to various actions, but when they are removed it returns to its original condition. On the contrary, if we fuse it with sulphur a chemical change takes place, a new substance is formed, and the iron does not return to its original condition. This idea of *change* is the fundamental chemical conception. The first man who made glass, or extracted a metal from its ore, effected a chemical change; the idea became most sovereign and dominant in alchemy, the attempt to change base metals into gold; it reigned throughout the period of phlogistic chemistry, for was not phlogiston a subtle entity which effected changes in matter according as it was assimilated by matter or rejected from it? It is equally the character of the chemistry of Lavoisier and Cavendish, of Davy and Dalton, of Berthelot and Cannizzaro.

The "philosopher's stone" (of which much more anon) was a substance supposed to *change* all things into gold; the "elixir vitæ" was a substance which was to *change* old men into youths; the "universal solvent" was to *change* everything to a liquid form. Let us look at some of the definitions of chemistry. Boerhaave says : " Chemistry is an art which teaches the manner of performing certain physical operations, whereby bodies cognizable to the senses, or capable of being rendered cognizable, and of being contained in vessels, are so changed by means of proper instruments, as to produce certain determined effects, and at the same time discover the causes thereof, for the service of various arts." Sir Humphrey Davy writes as follows :—" Most of the substances belonging to our globe are constantly undergoing alterations in sensible quantities, and one variety of matter becomes, as it were, transmuted into another. Such changes, whether natural or artificial, whether slowly or rapidly performed, are called chemical ; thus the gradual and almost imperceptible decay of the leaves and branches of a fallen tree exposed to the atmosphere, and the rapid combustion of wood in our fires, are both chemical operations. The object of chemical philosophy is to ascertain the causes of all phenomena of this kind, and to discover the laws by which they are governed." Quite recently Dr. Miller has defined chemistry as " the science which teaches us the composition of bodies," and such knowledge we can only obtain by pulling matter to pieces (analysis), or by building it up (synthesis). Dr. Hofmann of Berlin has defined the vast body of so-called organic chemistry as " the history of the migrations of carbon," and is not migration change of place ?

Let us then define chemistry as the science which treats of the various kinds of matter, whether simple or compound, of which the world is composed, their properties, and the laws which govern their combination with, and separation from, each other.

We will now discuss any ideas of the ancients which bear upon changed matter in any form or condition : thus, their early cosmogonies ; the knowledge they possessed of metals and compound bodies ; and their various technical operations, such as glass-making and smelting, alike demand our attention.

CHAPTER II.

Early ideas relative to the formation of the World—Thales of Miletus: Later beliefs in his doctrine—Anaximenes—Empedokles—Herakleitos—Anaxagoras—Demokritos—The Atomic Theory—Aristotle—The Ethereal Medium—Transmutation of the Elements—The Four-element Theory—Mode of interpreting it—Cause of the absence of Natural Science among the Ancients.

IF we compare all the earliest ideas as to the formation of the world, we find them resolve themselves into the belief that the ether and chaos, mind and matter, were the original principles of things. The ether, a subtle vivifying principle, "passing as a mighty breath over the chaos ; the chaos a boundless watery expanse without form." It was thus according to Sanchoniathon in the belief of the Phœnicians, and the twenty-five principles of the Hindu philosophy of San'chya are finally reduced to these—matter and spirit, nature and soul. The Egyptian deity was called Nûm as the spirit moving over the face of the waters, Pthah as the principle of production. The Hindu deity Brahme typified the productive force of nature. Among more western nations, Gaia, the personification of earth, was held to be the first that sprung from Chaos, and the wife of Ouranos.

Okeanos was their son, and according to Homer was
the source of all the gods. The worship of the
elements, and of the sun and moon, was among the
very earliest forms of worship; thus we have in India,
Agni the god of fire, Indra the god of the firmament;
the sun was sometimes worshipped as a symbol of
the deity, sometimes as a deity; fire was worshipped
by the ancient Persians as a symbol of the deity;
in the Homeric religion we find the Olympian Zeus,
lord of the air, who possesses absolute and universal
power. We must notice, too, Aïdoneus, the brother
of Zeus, and lord of the Underworld, said by some of
the Greek philosophers to designate earth, and undoubtedly an old nature power. Again, "Hephais-
tos," says Mr. Gladstone, "bears in Homer the
double stamp of a Nature Power, representing the
element of fire, and of an anthropomorphic deity who
is the god of art at a period when the only fine art
known was in works of metal produced by the aid of
fire." He is also one of the seven star-deities of
Chaldæa, the symbols and names of which were given
at an early date to the seven metals.

The elements of the Greek philosophers were, as
we shall presently show, rather *principles* than elements in the sense in which we speak of the sixty-
five elements now known to chemistry. There was a
marked tendency in the earliest period of Greek
philosophy to make one element or principle fundamental, and to evolve the other elements and the
world from it. Thales of Miletus, who lived in the
sixth century B.C., and who was called "the first of
natural philosophers" by Tertullian, and "the first

who inquired after natural causes" by Lactantius, affirmed that water was the first principle of things, perhaps, according to some writers, because Homer had made Okeanos the source of the gods. At least we are reminded of the boundless watery chaos of older cosmogonies. This doctrine of Thales was not without its supporters during the Middle Ages, and, indeed, the convertibility of water into earth and air was not absolutely disproved until about a century ago. One of the ablest supporters of the dogma was Van Helmont (b. 1577, d. 1644), who affirmed that all metals, and even rocks, may be resolved into water ; animal substances are produced from it, because fish live upon it ; and vegetable substances may be also produced from it. This last assertion he endeavoured to prove by what would appear to be a very conclusive experiment in those days, when neither the composition of the air, nor of water was known. He took a willow which weighed five pounds, and planted it in two hundred pounds of earth, which he had previously carefully dried in an oven. The willow was frequently watered, and at the end of five years he pulled it up and found that its weight amounted to one hundred and sixty-nine pounds and three ounces. The earth was again dried, and was found to have lost only two ounces. Thus it appeared that 164 lbs. of wood, bark, roots, leaves, &c., had been produced from water alone. Hence he in- ferred that all vegetables are produced from water alone ; not knowing, as was afterwards proved by Priestley, that a constituent of the atmosphere called carbonic acid had furnished the solid part of the tree, although, indeed, there was much water with it. Boerhaave devotes a page of his big book to a discussion of "whether water be convertible into earth." He concludes that the small earthy deposit observed when rain-water is distilled, arises from the particles of dust which had settled on the water before its introduction into the distilling vessel. Mr. Boyle had previously affirmed that "a very ingenious person, who had tried various experiments on rain-water, put him beyond all doubt about this transmutation, for he solemnly affirmed, on experience, that rain-water, even after distillation in very clean glasses, near two hundred times, afforded him this white earth." Finally, Lavoisier, in 1770, communicated to the *Académie des Sciences* an elaborate paper, "on the nature of water, and the experiments by which it has been attempted to prove the possibility of changing it into earth." In this he conclusively proved that water cannot be changed into earth, although it be distilled backwards and forwards for many successive days. Here then we find the old Thalesian theory at last disproved, but not before it had endured for twenty-four centuries; and this is by no means a solitary example of the permanence of old ideas. We shall become acquainted with yet older theories, which are still admitted, and which seem to be essential to physical philosophy.

On the other hand, Anaximenes regarded air as the primal element, Herakleitos fire, Pherekides earth, and some philosophers grouped two elements together. Anaximenes held that clouds are caused by the condensation of air, rain by the condensation of clouds; he appears to have clearly connected condensation with cold, rarefaction with heat. Archelaus affirmed that air when rarefied becomes fire, when condensed, water. It was very generally believed during the Middle Ages that water when boiled was converted into air. Empedokles introduced the idea of four distinct elements—earth, air, fire, and water, not capable of passing one into the other, but forming all things by their intermixture. These elements are acted upon by two principles, a uniting force of amity, a separating force of discord, corresponding somewhat to our attraction and repulsion. He endeavoured to prove the four-element theory by the following crude experiment: wood is burnt upon a hearth, fire seems to be evolved from it, the smoke is air, moisture is deposited on the hearthstone, while the ashes are earth:—hence wood is made up of earth, air, fire, and water. Empedokles was one of the first to materialise the Homeric gods; he applied his four-element theory even to them, declaring that Zeus was the element of fire, Here the element of air, Nestis the element of water, and Aidoneus the element of earth. Herakleitos (about 460 B.C.) made fire the primal element, and assumed that it condensed itself into the material elements, and that air, water, and earth were respectively formed as the fire became more condensed. He asserted, moreover, that all things are in perpetual motion and change, the moving force being fire; "fire is to him," says Schwegler, "even in individual things, the principle of movement, of physical as of spiritual vitality; the soul itself is a fiery vapour." We find in the fire of Herakleitos to some extent the attributes of what we now call a physical force; thus it is precedent to matter, and is the motive power of the universe; it influences and changes matter; it is perpetually undergoing transformation, but ultimately returns to its own form. Prof. Max Müller speaks of Herakleitos as "one of the boldest thinkers of ancient Greece." We can well understand why fire should, at a very early date, be regarded as chief of the elements, and the motive power of the universe; it had long been worshipped as a symbol of the deity by the Chaldæans; a worship which possibly originated with the Scyths; for Zoroaster, who introduced fire-worship among the Medo-Persic races, is supposed to have been a Scythian. Again, Agni, the god of light and_fire, was placed first in the Hindu Trinity.

Anaxagoras of Klazomene (500 B.C.) asserted that originally all things existed in infinite disorder; before the Creation there was a chaos of mingled particles of matter, which were arranged in order by a designing intelligence, or mover of matter, (*νοῦς*). The primitive constituents of things are not definite elements, like those of Empedokles, but are *homœomeries* (ὁμοιομέρειαι) that is *like parts*, small particles of matter like the masses they produce when they aggregate. Thus a mass of iron is produced by the aggregation of an infinite number of iron-homœomeries brought out of the chaos by the *νοῦς*, which latter possesses vortical motion which enables it to separate like parts and bring them together, somewhat on the principle of gold-washing.

If a dish containing substances of different relative weight, such as cork, sand, and lead shot, intimately mixed together, be caused to rotate, like particles will come together, the lead in one place, the sand in another, and this experiment will help us to realise to some extent the meaning of Anaxagoras when he assumes that the vortical motion of the νοῦς caused homœomeries to aggregate and form the world. Leukippos taught that the world is produced by the falling together of small indivisible particles or *atoms* (from ἀ and τέμνω), which are the principles of things, and which possess rapid circular motion. Demokritos (460 B.C.) extended the atomic theory of Leukippos ; he contended that the principles of things are atoms and a vacuum. The atoms are invisible by reason of their smallness, indivisible by reason of their solidity, impenetrable and unalterable. They have no other qualities, neither heat, nor cold, nor colour. Atoms are infinite in number, the vacuum is infinite in magnitude. Atoms differ from each other in size, shape, and weight. They are actuated by necessity or fate (ἀνάγκη), and possess an oblique motion in the vacuum which causes atoms of like shape to collide, and group themselves together, by which means all things are formed. The vacuum is necessary, otherwise motion of the atoms would be impossible, because there would be no place to receive them. Long before the time of Demokritos an atomic theory had been proposed in India by Kanáda, the founder of the Nyaya system of philosophy, of which this theory forms the distinguishing feature. The theory of Leukippos is

C 2

attributed by Possidonius to Moschus, a Phœnician. During the Middle Ages many writers made the atomic theory a prominent part of their system. Descartes adopted it in a somewhat modified form, and associated with his particles the vortical motion possessed by the homœomeries of Anaxagoras. Finally, almost in our own day, the atomic theory was introduced into chemistry by Dalton, and its introduction marked an important era in the science. At the present time the doctrine of atoms forms a principal feature in chemistry, and other branches of science find the conception most conducive to the philosophical explanation of phenomena. The definition of an atom given by Demokritos is almost as absolute and precise as that which we find in our most modern treatises. Thus the theory has endured for more than twenty-five centuries, and is likely to endure until there shall be no more science. It offers a striking example of the oneness of physical thought; the conception seems to be essential to Natural Philosophy; the most stupendous phenomena may be referred to atomic motions. S. Augustine has well said, "Deus est magnus in magnis, maximus autem in minimis."

The Hindus not only possessed the idea of the atomic constitution of matter, but further associated an attractive force with the atoms. This is well shown in the following extract given by Sir William Jones, from the poem of "Shi'ri'n and Ferhád, or the Divine Spirit, and a human soul disinterestedly pious":—"There is a strong propensity, which dances through every atom, and attracts the minutest particle to some peculiar object; search this Universe from its base to its summit, from fire to air, from water to earth, from all below the moon to all above the celestial spheres, and thou wilt not find a corpuscle destitute of that natural attractibility; the very point of the first thread in this apparently tangled skein is no other than such a principle of attraction, and all principles beside are void of a real basis; from such a propensity arises every motion perceived in heavenly or in terrestrial bodies; it is a disposition to be attracted, which taught hard steel to rush from its place and rivet itself on the magnet; it is the same disposition which impels the light straw to attach itself firmly to the amber; it is this quality which gives every substance in nature a tendency towards another, and an inclination forcibly directed to a determinate point."

The most prolific writer on Science amongst the ancients was Aristotle (b. 385, d. 322 B.C.). He was the author of various treatises, on the Heavens, on Generation and Corruption, on Physics, on Respiration, on Audibles, &c., and his views as well on metaphysics and ethics, as on science, were nearly universally accepted during the Middle Ages. Indeed, the scientific writings of Aristotle influenced science for nearly twenty centuries. Few, however, of his opinions concern us here. He was the first to introduce into Greek philosophy the *ether*, which he regarded as a fifth element (hence afterwards called *quinta essentia*) more subtle and divine than the other elements. The word quintessence is frequently used by the alchemists and early chemists, and is found in our most recent English dictionaries. The idea of an infinitely rarified and all-penetrating matter had long existed in physical philosophy, notably in the Hindu systems; it was probably recognised as a fifth element prior to the ninth century B.C. Aristotle is said to have called it αἰθήρ from ἀεὶ and θέω, because he conceived it to be always in motion, and to be the moving agency of the other elements; but we cannot admit this derivation now, and prefer to trace it to αἴθω and indh. In the present day we find it impossible to explain various phenomena, notably those connected with radiant heat and the polarisation of light, without assuming the existence of some rare ethereal medium, cubic miles of which would not weigh a milligramme, and we still call it the *ether*. Few physical systems have avoided this supposition; we make less use of it in chemistry than in physics; but it would be difficult to account for such actions as the combination of chlorine and hydrogen under the influence of light, without it.

Aristotle held that the four elements are mutually convertible, and he assigned two qualities to each, one of which was common to some other element. Thus he said:—

> " Fire is hot and dry.
> Air is hot and moist.
> Water is cold and moist.
> Earth is cold and dry."

In each of these one quality is dominant. Thus fire is more hot than dry, air more moist than hot, water more cold than moist, and earth more dry than cold. If the dry of fire be vanquished by the moist of water, air will result ; if the hot of air be vanquished by the cold of earth, water will result ; if the moist of water be vanquished by the dry of fire, earth will result. This idea of the

FIG. 1.—Alchemical Representation of the Transmutation of the Elements.

transmutation of the elements was adopted generally in works on alchemy ; the above figure, which embodies it, is from a work entitled "Preciosa Margarita Novella," published in Venice in 1546.

Aristotle's method of expressing the transmutation of the elements does not seem to differ much from that of earlier philosophers ; it would appear that he means to imply that if water be heated air is produced, while if it be heated more strongly so as to evaporate it to dryness, earth is left. His account of the generation of fire from air and earth is based on the most shallow and meagre observation, and shows to what results the most astute mind may be led if unaided by experiment. The generation of fire, he says, is made evident by the senses, for flame is notably fire, but flame is burning smoke, and smoke is from air and earth.

It is not here that we may tell how the philosophy of Aristotle was introduced into Europe by the Arabians, how from it arose that stupendous mass of false philosophy and perverted Aristotelianism called Scholasticism, and how for centuries the blind acceptance of the Peripatetic dogmas retarded the progress of science. Worse than all, Averroës, who has been called "l'âme d'Aristote," and who scattered Aristotelianism broadcast over Europe, did not know Greek, and the Latin versions of Averroës were " Latin translations from an Hebrew version of an Arabic commentary on an Arabic translation of a Syriac version of a Greek text." We may not, therefore, blame Aristotle for the results which followed from the too general and literal acceptance of his philosophy. Mr. Lewes has well said, "However he may have been impelled to systematise on imperfect bases, and to reason where he should have observed, it is not too much to say that had he reappeared

among later generations, he would have been the
first to repudiate the servility of his followers, the
first to point out the inanity of Scholasticism. His
mighty and eminently inquiring intellect would have
been the first to welcome and to extend the new
discoveries. He would have sided with Galileo and
Bacon against the Aristotelians."

We have spoken above of the endurance of the
Thalesian theory, that all things are formed from
water, and of the yet older theories of the existence
of an ethereal medium, and of atoms; but the
theory which affirms that the world is composed
of the four elements—earth, air, fire, and water,
is yet older, and is, indeed, the oldest physical
theory of which we have any knowledge. It certainly
existed before the fifteenth century B.C., it was
adopted in India, Egypt, and, as we have seen, in
Greece at a very early date. Then in the case of those
philosophers who made water, air, fire, &c., primal
elements, this element was first transmuted into the
three other elements, and the world was formed from
the four. We must be careful, however, to remember
that these four elements are not to be understood too
literally, they were rather principles or types of
qualities than actual elements. Several philosophers
divided fire into a purer and grosser part. Seneca
tells us that the Egyptians extended the theory by
assigning to each element an active and a passive
form: thus fire was divided into light which shines,
and into fire; air into passive atmosphere and active
wind; water into fresh and salt water; and earth
into cultivable land on the one hand, and rocks on

the other. These elements were extended yet more.
In later times *Fire* would come to signify everything appertaining to ignition; thus light, whether
accompanied by heat or otherwise, flame, the heat
inherent in all bodies, incandescent bodies, stars,
fiery meteors, lightning, and all visible manifestations
of electricity, would be included under the term.
Air would include smoke, steam, all vapours, and
whatsoever approached to the nature of a gas. When
gases were first discovered a hundred years ago,
they were called *Airs;* thus we read of *fixed air,
nitrous air, dephlogisticated air*, &c. *Water* would
include all liquids, of which, no doubt, blood, milk,
wine, and oil, were in early times the most familiar;
the words *aqua fortis, aqua regia, aguardiente, eau-ac-
vie,* &c., are vestiges of the old practice. *Earth*
included all rocks, however dissimilar they might be,
all kinds of cultivable land, metals, and whatever
appertained to solidity. Every solid was regarded
as a kind of earth at first. A century ago many
substances were called earths. At the present time
out of the sixty-five elements known to the chemist,
eight are classed as "earths" and three as "alkaline
earths." The fact is, the four ancient elements
were types of great classes of which the whole world
was constituted. In their most general sense, *earth,
water, air,* signified *solidity, liquidity, gaseity,* while
fire was the force exercising itself upon matter.
We have seen that the elemental fire of Herakleitos
is the mover of matter, the principle of movement,
that which produces perpetual changes around us.
Fire was the ψυχή, the anima, the soul, the vivifying

spirit. The mythological side of the belief is seen in the story of Prometheus, who is fabled to have stolen fire from Heaven and therewith vivified mankind. The philosophical side of the belief is seen in the dogmas of Herakleitos. The four-element theory evolved itself from the rude ideas about ether and chaos, mind and matter, before discussed; it is one of those crude physical theories which is enunciated and accepted by races the most diverse in character, country, faith, destiny. There is great oneness in the human mind in the matter of broad principles in crude cosmical ideas. And let us not forget that the four-element theory was universally accepted during the Middle Ages, and was only disproved a century ago, when air was proved to be a mixture of two gases, water a combination of two gases, fire the result of intense chemical action, and earth a mixture of some dozens of elementary bodies, some combined, some single. We do not deny that during the continuance of the four-element theory it may often have been taken in its strictly literal sense; but we do venture to assert that the richer and more cultured intellects regarded it in the light we have above described.

We can quite understand why there was so little natural science among the ancients, when we remember the absence of all experimental method and means, and the obstacle presented by the habit of mind which induced them to apply reasoning in place of experiment in the study of nature, to reason upon an immature or ill-observed fact, and to generalise upon altogether insufficient data. A

simple sophistry applied to observation could lead
to the most monstrous results. Take, for example,
the argument of Diodorus, as given by Sextus
Empiricus to prove that nothing is moved :—"If
a thing be moved, it is either moved in the place
where it is, or in the place where it is not. But not
in that wherein it is, because it rests in the place
wherein it is; neither in that wherein it is not, for
where a thing is not, it can neither act nor suffer.
Therefore nothing is moved." Again, Sokrates and
many of his followers taught that it was unwise to
leave those affairs which directly concern man, to
study those which are beyond his control and
external to him. Thus, to inquire into the nature .
and distance of the stars seems an useless speculation,
because even if we could ascertain these things,
we could neither alter the course of the stars nor
apply them to any benefit of mankind.

We have, however, seen above that many of the
Greek philosophers had more or less definite notions
concerning matter and force, and that they frequently
insist upon the transmutation of matter from one
form into another; so far and so far only are we
concerned with their dogmas in our inquiry into
the Birth of Chemistry. But we must not fail to
notice the existence at a very early date of the four-
element theory, of an atomic theory, of the idea
of an ethereal medium, of the idea of transforming
one kind of matter into another by the agency of
some motive principle. Neither let us forget to note
the similarity of principles in diverse philosophies ;
thus the homœomeries of Anaxagoras and the atoms

of Leukippos are clearly related, so, too, are the νοῦς
of Anaxagoras, the ἀνάγκη of Demokritos, the actuating form of fire of Herakleitos, the moving ether
of Aristotle. The links which bind together ancient
and modern physical thought are strong and endur-
ing ; and, since they have lasted during the rise and
fall of many nations, and during the most profound
changes in the mode and tone of thought, it is
not unlikely that they will endure as long as the
chain itself.

CHAPTER III.

Practical Chemistry of the Ancients—Metallurgy: Gold, Silver,
Electrum, Copper, Bronze, Tin.

IN the preceding chapters we have discussed such
theories of the ancients as involve the conception of
change of matter (notably the assumed transmutation
of the elements), and which hence concern the early
history of chemistry. Having done with theory, we
now have to inquire to what extent the ancients were
acqainted with practical chemistry, what metals or
other elements were known to them, and what processes
dependent upon chemical action. We do not, of
course, use the term "practical chemistry" strictly in
its present sense, because chemistry as a science was
altogether unknown to the ancients. Some have,
indeed, endeavoured to prove that the Egyptians must
have been acquainted with the science, from the skill
with which they used various metallic oxides for colouring glass; but we have no proof of this. Neither
Herodotus, nor Pliny, nor Vitruvius, indicates any
knowledge of chemistry, as a science, among either
Egyptians, Greeks, or Romans. Pliny, in his celebrated "Natural History," has laboriously amassed

all the practical science and pseudo-science which
the ancients possessed, and we find no mention of
either chemistry or alchemy. At the same time it is
impossible that the Egyptians and Sidonians can
have attained their marvellous skill in the manufacture and colouring of glass, and in the extraction
and working of metals, without the acquirement of a
considerable amount of knowledge of the properties of
matter, and of certain chemical changes. But this
knowledge could never be worked up into a comprehensive system; it resulted from the labour of
artizans, and the gulf between the philosopher and the
manipulator was both wide and deep. There could
be no union of practice and theory. Between Herakleitos with his theory that fire is the primal element,
the actuating force of the Universe, and the man who
wrought metals never so deftly, who applied fire to
the use and service of mankind, there was no sym-
pathy, no reciprocal transference of ideas. To reason
concerning the properties of matter with one's eyes
shut was all very well, but to experiment with matter,
to endeavour to determine the cause of such and such
a change by experiment, was utterly unworthy of a
philosopher. Anaxagoras is said to have made an
experiment to prove that there is no vacuum. Aristotle found that a bladder of air weighed in air
weighed more than the empty bladder (which if the
experiment be properly made, is by no means the
case), and hence concluded that the air has weight.
But these are solitary exceptions; the way to study
Nature, if she is to be studied at all, is, they main-
tained, to apply the pure, unaided intellect to the

study, and to keep mind and matter as distinct as possible. From all this it resulted that your workers in metals and in curious arts, your makers of glass and pigments, kept their knowledge of matter to themselves, as secrets to be handed down from father to son.

Seven metals were known to the ancients, viz., gold, silver, copper, tin, iron, lead, and mercury. The first six are mentioned by Homer, and appear to have been known from remote antiquity, while mercury was not known till a later date; it was, however, common in the first century B.C. The Greek word μέταλλον, whence *metallum* and *metal*, signifies *a mine*, hence it was applied to anything found in mines, notably metals; μέταλλον is connected with μεταλλάω, "to search for diligently."

Gold has been valued from the earliest ages, on account of the peculiarity of its colour, its lustre, and its unalterability in air. The metal is invariably found in the native state, that is, uncombined with other substances, hence no metallurgical operation is necessary for its extraction. It is very often met with in surface deposits, and in early times was undoubtedly far more common in alluvium and the beds of rivers than now. It would thus be easily extracted by washing, and the grains could readily be fused together into a mass. Gold mines formerly existed in Ethiopia, in which the gold was found in a matrix of quartz, like much of the Australian gold of the present day. These mines were worked by the Egyptians, who employed large gangs of slaves for the purpose. The quartz was crushed, and the

gold obtained from it by washing. We find representations of gold washings, and the subsequent
fusion of the metal, on Egyptian tombs, at least as
early as 2500 B.C., that is to say, about the time of
Joseph in Hebrew history. The woodcut (Fig. 2) is
given by Sir Gardner Wilkinson, and is taken from

FIG. 2.—Gold Washing: Fusion and Weighing of the Metal, from early
Egyptian Tomb.

a tomb at Beni Hassan. It represents gold washing,
and the fusion and weighing of the metal.

It is obvious that the process is only indicated, and
not accurately or minutely portrayed. Another form
of furnace is depicted in Fig. 3, and a blowpipe
somewhat different from that shown in Fig. 2. The
raised portion of the furnace is doubtless for the

D

purpose of concentrating the heat upon the crucible, on the principle of the reverberatory furnace.

FIG. 3.—Furnace and Blowpipe from Egyptian Tomb.

Gold once obtained was soon made into ornaments, very fine gold wire was used by the Egyptians for embroidery 3,300 years ago. Many of the Egyptian and Etruscan gold ornaments are very beautiful ; we may notice particularly the gold myrtle wreath found in an Etruscan tomb a few years ago. The Egyptians also used gold for inlaying, and it was beaten into leaf and used for gilding as early as 2000 B.C. In the Odyssey the gilding of the horns of an ox about to be sacrificed is mentioned.

Silver like gold, is often found native, and from several of its ores, the metal may be extracted by the action of heat alone. It has been known from the earliest ages, and was used chiefly for ornaments and embroidery. Gold was used for money before silver, which was first known as " white gold." The oldest silver Greek coin is a coin of Ægina, and was, perhaps, coined in the eighth century B.C. But the oldest coins in existence are the *electrum* staters of Lydia. Electrum consists of about three parts of gold to one of silver. Probably the metals were first found in nature thus alloyed, and as no method of separating them was then known, they were worked up together. Electrum was so called from its resemblance as regards colour to amber (ἤλεκτρον), which received its name from ἠλέκτωρ, the sun. It will be remembered incidentally that the science of Electricity was so called by Gilbert of Colchester, because the attractive force was first observed in amber. Amber is mentioned more than once by Homer. Electrum as a metal is first mentioned in the Antigone of Sophocles. It was found naturally alloyed, as in the pale gold of the Pactolus, which contains a good deal of silver; and was also made artificially. Probably all very pale gold was called electrum; Pliny states that gold containing a fifth part of silver is called electrum. In the British Museum there are many coins made of this alloy.

Copper was in use before iron. It is, as is well known, usual to denote various early ages by the substances then used for domestic implements. Thus we have the "age of stone," the "age of iron," &c. The stone age is followed by the age of copper, this by the age of bronze, and the age of bronze by the age of iron. Homer wrote in the age of copper; the shield of Achilles is made of gold, silver, tin, and copper; the arms and implements and utensils of his heroes are of copper. Mr. Gladstone has argued at some length that by chalcos (χαλκός) Homer meant copper, not bronze, as it is so often rendered. Chalcos is spoken of as a cheap and common metal, while tin was very scarce and rare; and it is scarcely probable

that so many things, even down to the commoner
utensils, could have contained ten or twelve per
cent. of tin. Again, Mr. Gladstone points out that
Homer speaks of chalcos as ἐρυθρός, red, a term
that could not apply to bronze; and he goes so
far as to say; "If chalcos be not copper, then copper is never mentioned in Homer" (*Juventus Mundi*,
p. 530). At the same time we must remember that
copper is very soft for cutting-instruments, and a
small quantity of tin hardens it. Some of the Greek
bronzes only contain 1 per cent. of tin. Dr. Percy
found in a bronze bowl of great antiquity from
Nineveh, copper 99·51, tin ·63. Ancient nails have
been found containing copper 97·75, tin 2·25; and
Mr. Gladstone suggests that, as tin is sometimes
found associated with copper in nature, this may
account for their composition. Copper is sometimes
found native, sometimes in the form of ores, from
which the metal is easily extracted. It appears to
have been both cheap and plentiful at an early date.
Romulus is said to have coined copper; it was also
used for money by the Egyptians. Great confusion
exists among old writers regarding the words signi-
fying bronze and copper; Pliny clearly did not understand the difference between copper and bronze.
The words *æs* and χαλκός appear to have been
applied indiscriminately both to copper and to alloys
of copper containing a large proportion of that metal.
Copper was alloyed with tin at an early date, because
copper is soft and is unsuitable for cutting-instruments,
while the addition of tin hardens it. The fusing
point of copper is between that of gold and silver, and

is far below that of iron, while the fusing point of tin
is only 446° Fahr. Thus the two metals could be
alloyed without any special metallurgical difficulties
or the requirement of an inordinate temperature.

Copper was first obtained by the Romans from
Cyprus, where it was very plentiful ; they called it
Æs Cyprium, which became corrupted into *Cuprum*,
from which we get our present chemical symbol for
copper, *Cu*. According to Solinus *æs* was found at
Chalkis, in Eubœa ; hence χαλκός, the Greek word for
copper. We read of "ores of æs," and of brass and
bronze being dug out of mines, whereas the term
brass is applied by us to an alloy composed of copper and zinc, and *bronze* to an alloy of copper and tin.
Zinc as a metal was unknown to the ancients, and
brass appears to have been made in Pliny's time by
heating together metallic copper, calamine (a native
carbonate of zinc), and charcoal ; the latter reduces
the calamine, and the metallic zinc and copper then
combine. According to Dr. Thomas Thomson,
aurichalcum or golden copper, was the proper name
for brass. *Æs* is to be always translated copper
or bronze, *not* brass, of which latter very little use
appears to have been made. Among other alloys of
copper, the ancients possessed the celebrated *Æs
Corinthiacum*, which according to Pliny was formed
accidentally during the burning of Corinth, by Mum-
mius, B.C. 146. There were four varieties of this, one
of which contained equal proportions of gold, silver,
and copper ; the others were most probably various
admixtures of copper and tin. The commonest kind
of ancient bronze contained in 100 parts, 88 parts of

copper, and 12 parts of tin. Two specimens of
bronze from Nineveh were found by Dr. Percy to
contain respectively—

	Bronze hook.	A small bell.
Copper	89·85	84·79
Tin	9·78	14·10
	99·63	98·89

The proportion of copper and tin (about 10 to 1)
is, remarks Mr. Layard, the composition of our best
modern bronze, while the increase of tin in the case
of the bell proves that the Assyrians were well acquainted with the increase of sonorousness produced
by changing the proportions of the metals. Modern
bell-metal contains about 80 parts of copper to 20
parts of tin. Sometimes a small quantity of lead
was introduced by the ancients into their bronzes.
Thus, a certain bronze for statues was formed by
fusing together 100 parts of copper, 10 parts of
lead, and 5 parts of tin. In a very ancient bronze
armlet (probably Phœnician) found in this country,
and belonging to a period anterior to the Roman
occupation, Prof. Church found—

Copper	86·49
Tin	6·76
Zinc	1·44
Lead	4·41
Oxygen and loss	·90
	100·00

Bronze was very much used in Egypt for vases,
mirrors, arms, &c. These, according to Sir G. Wilkinson, usually contain from 80 to 85 per cent. of
copper, with from 15 to 20 per cent. of tin. By the

use of some acid substance, the surface was sometimes covered with a green or brown patina. Although the casting of the metals was not known in Greece in the time of Homer, bronze was probably cast in Egypt 2000 years B.C. Several compounds of copper were used by the ancients : both the red and black oxide were obtained by heating copper to redness, and allowing it to cool in the air ; they distinguished between the scales which fell off during cooling, and those which were caused to fall off afterwards by the blows of a hammer. These oxides were principally used for colouring glass. Verdigris or acetate of copper was obtained then, as now, by covering plates of copper with the refuse of grapes after the expression of the vine-juice. Copper pyrites and a rude kind of sulphate of copper would appear from Pliny's obscure account to have been also known.

It follows from the above remarks concerning bronze, that tin, like copper, was known at a very early date. This is the more remarkable, because it has always been a comparatively scarce metal, and it was obtained from distant localities. Formerly it was almost entirely supplied by Spain and Britain. The Phœnicians, who were the earliest traders, obtained it first from India and Spain, and afterwards from Britain. The Greek name for tin, *kassiteros* (κασσίτερος),[1] was perhaps derived from the Insulæ

---

[1] The word κασσίτερος is used both by Homer and Hesiod, and it is possible that it may have been borrowed from the Sanskrit *kastíra*, and that tin was first procured from India. The Sanskrit word for tin, *kastíra*, is clearly related to the word *kás*, to shine. It is strange that the Arabic word for tin is *kàsdir*, closely resembling the Sanskrit,

Cassiterides, or Scilly Islands, from whence the
Phœnicians asserted that they procured tin ; but it
has been suggested that in all probability they invented the story because they desired a monopoly
of the metal, while in reality they procured all their
tin from the mainland of Cornwall, where it has
always abounded. Tin must have been very valuable
or the Phœnicians would not have traded so far for
it. Homer evidently considers it of far greater value
than copper. In the time of Pliny it was worth
about eight shillings the pound. The metal was
known in Egypt 2000 B.C. Pliny mentions that it
was found in the form of small black grains in
alluvial soils, from which it was obtained by washing ;
this account would agree with a description of the
so-called *stream tin*, which is tin ore separated from
the parent vein, and carried down by streams. It is
an oxide of tin, and the metal is obtained from it
by strong ignition with charcoal. Tin was used for
tinning copper vessels, for making mirrors, and in the
manufacture of bronze. In the Iliad the greaves of
the armour of Achilles are made of tin, and it enters
into the composition of the shield ; it was also used
for coating copper.

although there is no family relationship between the languages. Possibly the Phœnicians first procured tin from India, and gave it a name
resembling its native name *kastira ;* then the Greeks converted the
Phœnician word into κασσίτερος, the Romans borrowed the word from
the Greeks, and the fact of the scarce metal being found in certain
islands north of Spain, was sufficient to secure for them the distinctive
title of *Insulæ cassiterides*, or Tin Islands.

CHAPTER IV.

Iron — Lead — Quicksilver — Colours used for Painting and Dyeing—Glass—Certain Minerals known to the Ancients—Miscellaneous Processes.

IRON was not in common use till long after the introduction of copper. It is far more difficult to procure, because it is not met with in the native state, and the fusing point is very high. The metallurgy of iron is more complex than that of copper, and when obtained it is a more difficult metal to work. According to Xenophon the melting of iron ore was first practised by the Chalubes, a nation dwelling near the Black Sea, hence the name Chalups (χάλυψ) used for steel, and hence our word *Chalybeate* applied to a mineral water containing iron. Steel was known to the ancients, but we do not know by what means it was prepared; it was tempered by heating to redness, and plunging in cold water. According to some, kuanos (κύανος) mentioned by Homer was steel; but Mr. Gladstone prefers to conclude that it was bronze. Iron was known at least 1537 B.C. It was coined into money by the Lacedæmonians, and in the time of Lukourgos was in common use. It

was used in the time of Homer for certain cutting-
instruments, such as woodmen's axes, and for plough-
shares. Its value is shown by the fact that Achilles
proposed a ball of iron as a prize for the games in
honour of Patroklos. Neither iron money nor iron
implements of great antiquity have been found, because, unlike the other metals of which we have
spoken above, iron rusts rapidly, and comparatively
soon disappears. No remains of it have been found
in Egypt, yet Herodotus tells us that iron instruments
were used in building the pyramids ; moreover, steel
must have been employed to engrave the granite and
other hard rocks, massive pillars of which are often
found engraved most delicately from top to bottom
with hieroglyphics. Again, the beautifully engraved

FIG. 4.—Egyptian Bellows.   Fifteenth Century B.C.

Babylonian cylinders and Egyptian gems, frequently
of cornelian and onyx, must have required steel tools
of the finest temper. We have no record of the
furnaces in which iron ore was smelted, but we know
that bellows were in use in the 15th century B.C. in
Egypt, and some crucibles of the same period are

preserved in the Berlin Museum. They closely
resemble the crucibles in use in the present day.
The accompanying woodcut (Fig 4) represents a
double pair of bellows, a furnace, fuel, and perhaps a
crucible.

The native Indians prepare iron from hæmatite at
the present time by equally primitive bellows, which
indeed resemble the above very closely, and which,
without doubt, have been unaltered for centuries. A
small furnace, A (see the accompanying section, Fig.
5),[1] is rapidly constructed of clay, and into the bottom of this two nozzles, are introduced at B; these are

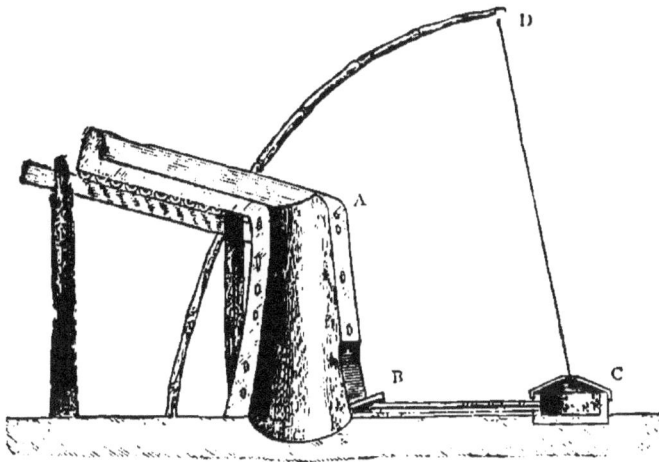

FIG. 5 —Smelting Furnace and Bellows used by native Indians in the present day.

connected with the bellows by bamboo tubes. The
bellows, C, consists of a cup-shaped bowl of wood
covered with goat-skin above, and connected with the

---

[1] We are indebted to Dr. Percy for permission to copy this figure
from his "Metallurgy," and to Mr. Murray for Figs. 2, 3, 4, and 7.

bamboo below.   In the centre of the goat-skin cover a
round hole is cut ; the blower places his heel upon this,
which is thus closed, while at the same time the skin
is depressed and a blast is driven from the tube, then
he steps upon the second skin, and thus a nearly con-
tinuous blast is kept up.   The bent bamboo and string,
D, is for the purpose of raising the goat-skin cover of
the bellows after depression, which, it will be noticed,
is accomplished in the Egyptian bellows by a string
raised by the hand.   A piece of hæmatite is introduced
with some charcoal, and after the lapse of some time,
it is reduced by the carbonic oxide to a spongy mass
of iron.   Undoubtedly a crude furnace and appliance
of this nature was used by the first smelters of iron.

Although we hear less of lead than of the preceding metals, it was known to the Egyptians at an
early date, and it is mentioned by Homer.   In the
time of Pliny leaden pipes were used to convey
water ; and sheet lead was employed for roofing purposes.   The chief supply of the metal came from
Spain and Britain.   Pliny believed that lead was
reproduced in the mine, so that if an exhausted mine
were closed it would be fit to work again in a few
years' time.   This idea of the growth of the metals
was very generally accepted by the alchemists.   Tin
and lead were sometimes alloyed together by the
ancients, and tin was used as a solder for lead.
Litharge or protoxide of lead, and *cerussa usta* (burnt
ceruss), or red lead, were used by painters.   *Cerussa*,
which we now call "white lead," or, more strictly, carbonate of lead, was prepared by exposing sheets of
lead to the fumes of vinegar in a warm place, a heap

of decomposing manure, for instance. A basic acetate of lead is formed by this means, which is partially converted into carbonate by the carbonic acid given off by the decomposing organic matter. Cerussa was used by Athenian ladies as a cosmetic. *Cerussa usta* was first formed accidentally from cerussa during the burning of a house near the Piræus. Litharge is easily formed by heating lead above its melting-point in air, when it absorbs oxygen gas, and the resulting oxide may be skimmed off.

Mercury was common in the time of Pliny, but it is not mentioned by earlier writers. It was found native in Spain, but was more generally obtained by heating cinnabar (sulphide of mercury) with iron filings in an earthen vessel, to the top of which a cover was luted. The iron decomposed the sulphide, and the liberated mercury was volatilized and condensed on the cover of the vessel, whence it was collected. This method, described by Dioscorides, is the first crude example of *distillation*, which afterwards became a principal operation among the alchemists and chemists for separating the volatile from the fixed. In the time of Dioscorides cinnabar was called *minium*, but it became so largely adulterated with red lead that the term minium was ultimately applied to the latter. Minium is still one of the names for red lead. Pliny was acquainted with the high specific gravity of mercury, and with its power of dissolving gold. Substances were sometimes gilded by a gold amalgam. Mercury was also used, as now, for extracting gold from its earthy matrix ; the gold-bearing rock was powdered and shaken up with

mercury, which dissolved out the gold ; the amalgam
of gold and mercury was then squeezed through
leather, which separated most of the mercury ; the
solid amalgam was heated to expel the mercury, and
pure gold remained. Vitruvius states that gold was
recovered from gold embroidery by burning the cloth
in an earthen pot, and throwing the ashes into water
to which quicksilver was added. The latter attracted
the gold and dissolved it ; the amalgam was put into a
piece of cloth and squeezed between the hands, and the
mercury, on account of its fluidity, was forced through
the pores of the cloth, while the gold remained.

Native mercury was called *argentum vivum* (quicksilver), while mercury distilled from cinnabar was called
hydrargyrum (ὕδωρ ἄργυρον, liquid silver), from which
we take our present symbol for the metal, *Hg*. The
alchemists, in whose eyes, as we shall hereafter see,
mercury was a very important metal, call it by the
various names of *mercurius, argentum vivum, hydrargyrum*, with others of a more fanciful nature.

The ancients were not acquainted with any other
metals in an uncombined state, except the seven
mentioned above. *Stibium*, or sulphide of antimony,
was used in the East at an early period for painting
the eyelashes. It is still used for that purpose, and
is called *kohl*. Native carbonate of zinc was known,
and black oxide of manganese. The two sulphides
of arsenic were known, and were used as pigments.
The yellow sulphide was called *auripigmentum* and
arsenicum ; the red sulphide went by the name of
sandaracha. Auripigmentum became contracted into
orpiment, a word which we find both in alchemical

treatises and in our most modern treatises on chemistry.

The colours used by the ancients for painting were examined by Sir Humphry Davy at the beginning of this century, and he came to the conclusion that "the Greek and Roman painters had almost all the same colours as those employed by the great Italian masters at the period of the revival of arts in Italy." Various colours have been examined from the frescoes in the Baths of Titus, from Pompeii, and from Egyptian tombs. The colours of the Egyptians were red, yellow, blue, green, black, and white. The red was bole, that is a clay deriving its colour from oxide of iron ; the yellow, an ochre, also clay, coloured by a paler form of oxide of iron ; the green, a mixture of this ochre with a blue powdered glass, produced by fusing together sand, carbonate of soda, and oxide of copper. The black was ivory black, prepared by heating bones out of contact with air until completely carbonized ; the white was powdered chalk. These various colours were mixed with gum and water before use. The Greeks and Romans used red lead and cinnabar, as well as red ochre, and yellow protoxide of lead. The blue powdered glass mentioned above was sometimes called κύανος by the Greeks, *Cæruleum* by the Romans. Vitruvius describes the method of preparing it ; and Davy prepared a substance which perfectly resembled the ancient colour, by fusing together fifteen parts of carbonate of soda, with twenty parts of powdered flints, and three parts of copper filings.

The green of the Romans was carbonate of copper, and for browns they sometimes used dark oxide of

manganese. The *purpurissum* of the Romans was
Tyrian purple, a very valuable colour obtained from
a shell-fish, and much used for dyeing. In order to
obtain the colour for the purposes of painting, clay
was placed in the chaldrons of dye, so as to absorb
the colour, and was afterwards removed and dried.
Indicum purpurissum was probably indigo ; Pliny
mentions that the vapour possesses a fine purple
colour. Ivory black was called *Elephantinum* ; lamp
black, that is soot, was called *Atramentum.* The
latter mixed with water constituted the ink of the
ancients.

According to Pliny, glass was first discovered by
some Phœnician merchants who were returning from
Egypt with a cargo of *natron* (carbonate of soda),
and who landed on the sandy banks of the river Belus.
In order to support the vessels they used for cooking
their food over the fire, they used some large lumps of
natron, and the fire was sufficiently strong to fuse it,
with the fine sand of the river. Hence resulted the
first glass. Whatever may be the value of this story,
we find representations of glass-blowing on the monu-
ments of Thebes and Beni Hassan ; and the Egyptians were well acquainted with it 2450 B.C. The
most celebrated manufactory of glass was in Egypt ;
and, according to Strabo, a peculiar kind of earth
found near Alexandria was essential for the finer kinds
of glass. The Egyptian glass had nearly the same
composition as our " crown glass," which contains 63
per cent. of silica, 22 of potash, 12 of lime, and 3 of
aluminia. The Phœnicians and Egyptians exported
large quantities of glass to Greece and Rome. The

Egyptians engraved and cut glass with the diamond ; they also possessed extraordinary skill in colouring glass with various metallic oxides, and combining several colours in the same vase, and they imitated precious stones with great success. We read of whole statues made of emerald, but these were undoubtedly of emerald glass, viz. glass coloured by oxide of copper. The Egyptians understood the art of enamelling on metal. Aristophanes is the first Greek author who mentions glass (τὴν ὕαλον) ; he alludes to the use of a lens of glass, as a burning-glass in the Νεφέλαι, which play was acted in Athens, B.C. 423. Colourless glass was the most valuable, and a small quantity of oxide of manganese was added then as now for the purpose of decolourising it. A very ancient opaque green glass, analysed by Klaproth, was found to contain 65 per cent. of silica, 10 of oxide of copper, 7·5 of oxide of lead, 3·5 of oxide of iron, and about 6 per cent. of both lime and alumina. A red glass was found to be coloured by red oxide of copper.

Dyeing was much practised by the ancients; the Egyptians understood the effect of acid on some colours, and were acquainted with mordants, that is, substances which "fix" the colouring matter in the fabric, and prevent it from being washed out. The most celebrated dye of antiquity was the purple of Tyre, discovered about 1500 B.C., perhaps earlier. It was produced by certain shell-fish which inhabit the Mediterranean ; these are spoken of as *buccinum* and *purpura* by Pliny. A few drops only of the dye were obtained from each fish, and the colour

E

hence became very valuable, and was monopolised by the emperors of the world. The Egyptians dyed linen with indigo, which they procured from India, for they had considerable intercourse with that country at an early period.

Lime was used for removing the hair from skins about to be tanned. Leather made in the time of Sheshonk, the contemporary of Solomon, has been found in a good state of preservation. For the process of tanning, they used the pods of the Acacia Nilotica, a plant which, according to Sir G. Wilkinson, was also prized for its timber, charcoal, and gum.

Nitrum was a term applied to carbonate of soda, or natron, which, we have already seen, was used in the manufacture of glass. The substance which we now call *nitre* (nitrate of potash) was probably known in India and China before the Christian era. Dr. Thomas Thomson has suggested that when the real nitre was imported into Europe, it received the same name as carbonate of soda (nitrum), from the similarity of its appearance, and retained the name on account of its greater importance. Roger Bacon always speaks of nitrate of potash as nitre. The low Latin name for soda became *natrium*, hence our present symbol for sodium, *Na*.

Soap is first mentioned by Pliny ; it was made by mixing wood ashes, which contain carbonate of soda, with animal fat. It was used solely as a kind of pomatum. The Greeks added wood ashes to water to increase its cleansing properties.

The only acid with which the ancients were acquainted was acetic acid or vinegar. It has been suggested that the Egyptians discovered nitric acid and nitrate of silver, because a silver stain has been found upon some linen, but the evidence is insufficient. We remember the story of Cleopatra dissolving two pearls, valued at ten millions of sestertii, in vinegar; although only a story, it would seem to show that vinegar was the most powerful solvent known. This is further indicated by the story of Hannibal dissolving rocks by vinegar.

A number of minerals are mentioned by Pliny, but we can recognise but few of them. Iron pyrites (sulphide of iron) was used for striking fire with steel in order to kindle tinder, and was hence called pyrites ($\pi\hat{\upsilon}\rho$, fire), or fire-stone. Sulphur was well known, and was used for matches; it was also apparently burnt in a current of air, and the sulphurous acid produced employed was for bleaching purposes. Asphalt was used for embalming, and undoubtedly also for torches.

CHAPTER V.

Association of the seven Metals with the seven greater Heavenly Bodies—Consequent introduction of Symbols into the history of Matter.

THUS far we have become acquainted with the various theories of the Ancients, in which changes in the composition of matter are discussed, and with various processes by which changes were actually effected. Before we leave this period, and pass at one bound to the eighth century A.D., we must notice the commencement of a symbolical system in the history of matter, which in the hands of the Alchemists and early Chemists assumed vast proportions, and still appertains to the science of Chemistry. This system was commenced by the association of the seven metals with the seven greater heavenly bodies. We do not know at what period the metals were designated by the names and symbols of the planets : certainly in a very remote age.

At a very early date the Chaldæans represented the stars by symbols, and these gradually increased until astrology became one mass of symbols. On the occasion of certain religious ceremonies the Kings of Assyria wore a necklace in which the sun, moon, and stars were represented as emblems, for they were first worshipped as emblems of the Deity. Sculptural representations of necklaces with seven discs upon them have also been found. Symbols were carried before Egyptian priests, and their gods were represented with certain signs symbolical of their special attributes. The Assyrian goddess Astarte carries in her left hand a symbol (*b*) (Fig 6) not very different

FIG. 6.—*a* Crux ansata of the Egyptians; *b* Assyrian symbol of Astarte; *c* Later symbol of the planet Venus.

from the *crux ansata* of the Egyptians (*a*) ; and the symbol (*c*) by which the planet Venus was afterwards represented by the astrologers, and is still represented by astronomers. In the celebrated "Book of the Dead" (B.C. 1350), the most perfectly preserved Egyptian ritual which the world possesses, this latter symbol (*c* in the figure) occurs frequently among the hieroglyphics. This is very noticeable in the "Judgment scene" of the Turin papyrus, a copy of which exists in the British Museum. The upper portion of the *crux ansata* was frequently made more rounded in form, and it is obvious that if, in addition to this, the cross was somewhat lowered, we should arrive at the third symbol (*c*) shown above. The *crux ansata* (*a*), if written quickly, could easily pass into this latter

symbol (c), and this may account for the occurrence
of both symbols in the judgment picture, to which we
have alluded above.

Plato speaks of the sun, moon, and five planets,
but does not distinguish them by the names of gods;
Epinomis mentions them in conjunction with the
names of gods. It is probable that the Chaldæans
also associated the principal heavenly bodies with the
names of deities—San with the sun, Hurki with the
moon, Bel-Merodach with Jupiter, Astarte or Ishtar
with Venus, Nergal with Mars, &c. The relative
position of the planets was generally as follows : the
Earth was the centre of the system; next in order
came the Moon, the Sun, Venus, Mercury, Mars,
Jupiter, and Saturn; but these positions were sometimes varied. It was known that Saturn completed a
revolution in about thirty years, while Jupiter required
twelve years, Mars only two, and Mercury and Venus
appeared to take about the same time as the Sun;
hence the above order. As Saturn was farthest from
the source of heat, and the slowest in his motion, he
was supposed to be of an icy character, and to assert
an evil influence.

While speaking of the seven greater heavenly
bodies, and the seven metals, we may allude incidentally to the curious prominence of that number
in many matters—"that mysterious number," as
Mr. Layard calls it, "so prevalent in the Sabæan
system." Thus (to select a few instances at random)
we have seven days of the week, seven wise men of
Greece, seven wonders of the world, seven cardinal
sins, seven-stringed lyre, seven harmonic proportions,

seven heavens, seven walls of Ecbatana, seven gates of Thebes. The list might be extended almost indefinitely. Among the Hebrews the number was specially prominent. Not to mention the frequent allusion to it in the Apocalypse, we may recall the incidents of the fall of Jericho: the town was surrounded for seven days; on the seventh day the walls fell at the blast of seven trumpets, which were carried round the walls seven times by seven priests.

We cannot tell why the seven metals were associated with the seven deified heavenly bodies, unless it was because all things which amounted to the same number were connected with them. This, at least, we know, that long before the time of Geber, the first writer on chemistry, the metals had received the same names and symbols as the planets. "There is abundant evidence," says Mr. Gladstone, "of a correspondence between the seven metals of Homer and the seven metals of the ancient planetary worship of the East." In the time of Homer only six simple metals were known, and the seventh was the compound *kuanos;* quicksilver afterwards became the seventh simple metal, and received the name and symbol of the seventh planet. The metals were apportioned as follows :—

Gold	The Sun	☉
Silver	The Moon	☽
Quicksilver	Mercury	☿
Copper	Venus	♀
Tin	Jupiter	♃
Iron	Mars	♂
Lead	Saturn	♄

Herodotus tells us that Ecbatana had seven walls, the outermost of which was the lowest, and the others gradually ascended like steps to the highest, which enclosed the king's palace. They were each painted of a particular colour; the outermost white, the second black, the third purple, the fourth blue, the fifth red, the sixth the colour of silver, the seventh the colour of gold. Undoubtedly these had reference to the seven greater heavenly bodies. It is impossible to account for the colours, but it is curious to notice the particular colour which would fall to any particular metal. Placing the planets in order as applied to the metals, we should have gold to gold, silver to silver, red to copper, blue to iron, purple to tin, black to lead, the most despised of the metals. It is probable that the Sabæans associated these colours with the seven heavenly bodies. The temple of Bel-Merodach, rebuilt by Nebuchadnezzar, and called by him the "Wonder of Borsippa," appears also to have consisted of seven terraces differently coloured. The following is a portion of the inscription from a clay cylinder found among the ruins of the temple:—"I (Nebuchadnezzar) have completed the magnificence of the tower with silver, gold, precious stones, enamelled bricks, fir, and pine. . . . . This most ancient monument of Borsippa is the house of the seven lights of the earth."

How the symbols conferred upon the planets and afterwards upon the metals arose it is difficult to say; they are undoubtedly of Chaldæan origin, but to what extent they have since been modified no one can tell. They exist in early MSS. on Alchemy.

That the sun should be represented by a circle, the symbol of perfection, is no wonder. Again, that the moon should be symbolized by a crescent we can understand ; but the others present greater difficulties. Among these, some say we have the looking-glass of Venus, the thunderbolts of Jupiter, the spear and shield of Mars, the scythe of Saturn, and the caduceus of Mercury. In the temple of Hermes at Pselcis, he is represented with a staff having a serpent twining around it, from which it has been suggested the caduceus of Mercury may have been derived. (See Fig. 7, p. 65.) Some see in ♃, not the thunderbolts, but the throne of Jupiter ; others the *Zeta* of Zeus ; others, again, the Arabic 4, indicating that Jupiter was the fourth planet in order. Some, too, have seen in ♄ the K of Kronos. It is less difficult to understand why a particular metal was assigned to a particular heavenly body. Thus gold would naturally be associated with the sun, on account of its colour, perfection, and beauty, and because it was ever regarded as the noblest metal. For the same reason silver would fall to the moon, with its pale, silvery colour and light. So, again, iron, the metal of war, would be associated with Mars ; lead, the dull, despised metal, with Saturn, the slowest of the planets; quicksilver, the nimble volatile metal, with Mercury, the messenger of the gods.

These signs became in the hands of the Alchemists the commencement of a symbolic system in chemistry.

CHAPTER VI.

The Alchemists —Origin of Alchemy—Hermes Trismegistus— Greek MSS. on Alchemy—Their probable authorship and age.

WE spoke in the last chapter of the alchemists almost for the first time, and we must now turn our attention to the origin and growth of their dogmas, and to their work. We have already seen that the word χημεία is first found in the Lexicon of Suidas, and that he defines it as "the preparation of gold and silver." He further tells us, under the same heading, that the books on the subject, were sought for by Dioclesian and burnt, lest the Egyptians should become rich through their knowledge of the art, and should thus be able to resist the Romans. Now, the people who professed a knowledge of the art of making gold were called *alchemists.* The word *alchemy,* as we have previously shown, consists of a Coptic root united with an Arabic prefix, and signifies the *hidden* or *obscure art.* Alchemists were those who practised this mysterious art. We can well understand why the professors of such an art should maintain the utmost secrecy;

to divulge such magic would be to make all men
equally rich ; hence it was necessarily a hidden art.
Neither did the books on the subject avail much, for
they are filled with some of the most incomprehensible
nonsense that ever was written. Yet the literature
of the subject is enormous. The volumes on alchemy
in our large libraries are to be counted by the
hundred. In 1602 Zetzner published, in Strasburg,
a "Theatrum Chemicum," containing more than
a hundred tracts on alchemy, selected from various
notable authors. A century later Mangetus published his "Bibliotheca Chemica Curiosa," in two
large folios, containing a hundred and twenty-two
alchemical treatises. We have previously given the
titles of a few Greek MSS. on alchemy. The list
has been extended to eighty-three. Arabic and
Persian MSS. on the subject are not uncommon.
There are treatises in Spanish, Italian, German,
Dutch, and English on alchemy, and, more numerous
than all, treatises in Latin, in every large library.
Let us endeavour to get from the tangled mazes
of this hieroglyphical literature some idea of alchemy,
and of its influence upon chemistry.

We are, perhaps, puzzled at the outset to comprehend how any one man, much less thousands
of men, could have deluded themselves with the
belief in the possibility of transmuting one kind of
matter into another :—crude lead, or tin, or mercury,
into weighty, lustrous gold. But this was not the
greatest wonder of the age. At the time when
alchemy arose, and throughout the period during
which it most flourished, the belief in theurgy,

witchcraft, necromancy, and magic of all kinds was
rife among all classes; and surely it was less wonderful to change lead or tin into gold, than to call
up the spirit of one's ancestor, or to confer perpetual
youth upon a nonagenarian! It is, for wonderment,
as compared with the greater magic of the day, as
the process for the conversion of benzine into aniline
compared with spirit-rapping; or as a demonstration
of specific inductive capacity compared with a mani-
festation of psychic force. Alchemy was considered
to be perfectly rational not two centuries ago, and
was among the lesser forms of magic, inasmuch as
it did not require the influence of supernatural
causes.

The growth of the idea is not difficult to trace.
The ancients had persistently asserted the change
of one element into another. Thales, as we have
seen, evolved the ten thousand forms of nature and
kinds of matter, from water, Anaximenes from air,
by successive transmutation. Aristotle, whose physical views were accepted without question by the
alchemists, had endeavoured to show by clever
argument that, if you transfer a quality of water
to fire, you obtain air; while if you transfer a
quality of earth to air, you get water; and so for
fire and earth, and that from these elements all
things proceed. This was readily accepted by
Middle Age thinkers. The alchemists reasoned,
plausibly enough:—if fire becomes air, air water,
and water earth, why may not one kind of substance
formed from these elements be changed into another .
kind of substance of somewhat the same nature,

and certainly more similar than air and water, or
water and earth? Why may not lead, compounded
of these elements in certain proportions, be changed
into gold, compounded of these elements in certain
other proportions? There have been falser modes
of reasoning than this in the history of science.
Let the ancient Greek theory of the transmutation
of the elements be once literally accepted, and the
alchemical belief in transmutation follows naturally;
it is a minor application of the major proposition.
There is nothing to wonder at in this; the human
mind seldom moves by fits and starts; an essentially
new mode of thought and new form of belief is rare,
and many apparently new dogmas are united with
older dogmas in the closest manner, and are in fact
direct emanations from them. Such was the alchemical idea of transmutation. Admitting the
possibility of the process, a man would naturally
ask himself "What do I most desire to make?"
"What in this world procures the greatest amount
of happiness and of power?" For what have men
slaughtered each other by the thousand in open
war, or singly and secretly in the dead of night?
For what have kingdoms been sold, great tracts of
land ceded, and people been ground into serfdom till
they rose and rioted against their oppressors? For
what have princes and cardinals been created, emperors and kings destroyed, and the eternal peace of
troubled souls promised? In a word, for what will
man dare all things, sacrifice all things; for what will
he toil during a lifetime; to what will he devote all
his intellectual energies? This is surely the thing

for the ready acquirement of which we may devote much time and thought, and this thing is *gold*. This is the key to the prodigious masses of alchemical literature, and to the mysteries and anomalies, connected with men who often wasted their whole lives and all they possessed in the endeavour to change baser metals into gold.

If we consult alchemical MSS., no matter the date or author, or language, we find constant mention of Hermes Trismegistus, who was indeed considered, and sometimes designated, the *father of alchemy*. In a treatise attributed to Albertus Magnus we are told that the tomb of Hermes was discovered by Alexander the Great, in a cave near Hebron. In this was found a slab of emerald which had been taken from the hands of the dead Hermes by Sarah, the wife of Abraham, and which had inscribed upon it in Phœnician characters the precepts of the great master concerning the art of making gold. The inscription consisted of thirteen sentences, and is to be found in numerous alchemical works. It is for the most part quite unintelligible, and in style closely resembles the great mass of Middle Age alchemical literature.

The following is cited as the inscription of the " Smaragdine Table," and is to be found in very early MSS. in various languages :—

1. I speak not fictitious things, but that which is certain and most true.

2. What is below is like that which is above, and what is above is like that which is below, to accomplish the miracles of one thing.

3. And as all things were produced by the one word of one Being, so all things were produced from this one thing by adaptation.

4. Its father is the sun, its mother the moon ; the wind carries it in its belly, its nurse is the earth.

5. It is the father of all perfection throughout the world.

6. The power is vigorous if it be changed into earth.

7. Separate the earth from the fire, the subtle from the gross, acting prudently and with judgment.

8. Ascend with the greatest sagacity from the earth to heaven, and then again descend to the earth, and unite together the powers of things superior and things inferior. Thus you will obtain the glory of the whole world, and obscurity.will fly far away from you.

9. This has more fortitude than fortitude itself; because it conquers every subtle thing and can penetrate every solid.

10. Thus was the world formed.

11. Hence proceed wonders, which are here established.

12. Therefore I am called Hermes Trismegistus, having three parts of the philosophy of the whole world.

13. That which I had to say concerning the operation of the sun is completed.

The story and the inscription, together with all books attributed to Hermes (who is asserted to have lived about 2,000 B.C.), are no doubt the production of monks of the Middle Ages. In spite

of the obvious worthlessness of the inscription of
the emerald table, men have not been wanting who
have laboured long and lovingly to prove its authenticity, to interpret it, and to show that it is in
good sooth a marvellous revelation, full of sublime
secrets of considerable import to mankind.

Hermes Trismegistus is generally asserted by the
alchemists to have been a priest who lived a little
after the time of Moses. According to Clemens
Alexandrinus he was the author of forty-two books
containing all the learning of the Egyptians; others
tell us that he was the author of several thousand
volumes. Plato speaks of him in the " Phædrus " as
the inventor of numbers and letters. He was in fact
the Egyptian god of letters, and as such of course
could be described as the author of multitudinous
works. He was the deified intellect, and hence has
often been confounded with Thoth, "the intellect."
Sir Gardner Wilkinson speaks of Hermes as an
emanation of Thoth, and as representing "the ab-
stract quality of the understanding." The woodcut
(Fig. 7) representing Hermes, is from a temple at
Pselcis, which was erected by Erganum, a contemporary of Ptolemy Philadelphus. It may be well to
note the extent of the symbolism associated with the
sculpture; in one hand Hermes holds the *Crux ansata*,
the symbol of life, in the other a staff associated with
which are a serpent, a scorpion, a hawk's head, and,
above all, a circle surrounded by an asp, each with its
special symbolical significance. On the Rosetta stone
Hermes is called "the great and great," or twice
great ; he was called *Trismegistus*, or thrice great,

according to the twelfth aphorism of the Emerald
Table, because he possessed three parts of the wisdom
of the whole world, which in his light of deified intellect he might well do.

FIG. 7.—Hermes Trismegistus; from the Temple at Pselcis.

Perhaps no author is more often quoted by the
alchemists than Hermes, the supposed father of their
art.   They called themselves *Hermetic philosophers.*
Alchemy is often called the *Hermetic Art*, or simply
Hermetics.   To enclose a substance very securely, as

F

by placing it in a glass tube and fusing, or sealing, the mouth of the tube, was called securing with "Hermes his seal," and the echo of the idea lives amongst us yet; for, in our most modern treatises, the expression "to seal hermetically" may be found. Petrus Hauboldus, of Copenhagen, was surely one of the most enterprising publishers of his day, for he had the temerity to publish a book entitled *Hermetis Ægyptiorum et Chemicorum Sapientia.* A book square as to its dimensions, small as to its type, drier than dust as to its contents, of four hundred odd pages, of two centuries of age, writ in Latin, with a sprinkling of contracted Greek, and floridly dedicated to Jean Baptiste Colbert. A book wherein the author endeavours to prove that alchemy was known before the flood, that Hermes Trismegîstus was a real personage, the inventor of all arts, the father of alchemy, and much else besides. We may well imagine that the author of such a treatise was no ordinary man, and our conjecture proves a tolerably correct one. Olaf Borch, whose Latinised name became the more resounding *Olaus Borrichius,* was apparently the great mainstay of the University of Copenhagen; at all events, he was simultaneously Professor of Philology, Poetry, Chemistry, and Botany; and we must either imagine that, in 1660, professors were difficult to procure in the Kingdom of Denmark, or else that Olaus Borrichius was such an astounding genius that he could readily undertake the duties of four diverse professorships at the same time. We can scarcely imagine three greater antitheses than the philological faculty, the poetical faculty and the chemical faculty;

but here we find them united, or assumed to be united, in one man. Yet more, Borrichius was appointed Court Physician, and Assessor of the Supreme Court of Law. He was the very personification of all learning, if we may judge by the treatment he received from his countrymen. In addition to the work mentioned above, he wrote various treatises on philology, on the quantity of syllables, on the Greek and Latin poets, on medicine, chemistry, and botany. It is strange that a man who, presumably in his capacity of judge, was in the habit of sifting evidence, and of avoiding hasty generalisation, should have endeavoured with much elaborate argument to prove that Hermes Trismegistus was a real personage ; that his Smaragdine table was really found by the wife of Abraham, and that it contained matter of the highest import to mankind. We must imagine that in this matter Borrichius allowed the imaginative faculty due to his poetical temperament to exert an undue influence over his more sober judgment. He is equally at pains to assert the authenticity and antiquity of the various Greek MSS. on alchemy in the libraries of Europe. He specially mentions a MS. by Zozimus of Panapolis, on the art of making gold, in the King's Library in Paris; and Scaliger tells us that this same MS. was written in the fifth century. M. Ferdinand Hoefer is apparently penetrated by the Borrichian spirit of faith and imagination, and he unhesitatingly accepts the early date attributed to the Paris MS.

M. Hoefer traces the rise of Alchemy to the fourth century of our era ; it was then known as the "sacred art" (*ars sacra; τέχνη ἱερά*), and one of the chief

F 2

writers on the subject was the said Zozimus of
Panapolis. The principal Greek MSS. attributed to
Zozimus, which exist in the Bibliothèque Nationale,
have the following titles :—(*a*) On Furnaces and
Chemical Instruments ; (*β*) On the Virtue and Composition of Waters ; (*γ*) On the Holy Water ; (*δ*) On
the Sacred Art of making Gold and Silver. In the
latter, Zozimus mentions that if the "soul of copper,"
which remains above the water of mercury, be heated,
it gives off an aëriform body ($\sigma\hat{\omega}\mu\alpha$ $\pi\nu\epsilon\nu\mu\alpha\tau\iota\kappa\acute{o}\nu$),
and this (says M. Hoefer) was probably oxygen gas,
while the soul of copper was oxide of mercury. A
second author of early Greek MSS. was Pelagius,
who alludes to two writers named Zozimus—one
the "Ancient," the other the "Physician." A third
author, Olympiodorus, who calls the "sacred art"
chemistry ($\chi\eta\mu\epsilon\acute{\iota}\alpha$), quotes Hermes, Democritus, and
Anaximander as alchemists.

Democritus (not to be confounded with the Greek
philosopher of that name), in his "Physics and
Mystics," informs us how he invoked the shade of
his master, Ostane the Mede, and how the spirit
appeared and accorded him mystical communings.
Synesius, the commentator of Democritus, lived, according to M. Hoefer, about fifty years after Zozimus
(say 450 A.D.) ; but a treatise on the Philosopher's
Stone is in existence which claims Synesius as its
author, which mentions Geber, who lived at least 400
years later. Mary the Jewess, who is often alluded
to by later alchemists, was a contemporary of Democritus, and a writer on alchemy ; she also invented
various chemical vessels, among others a bath, to

gently transmit heat by means of hot sand or cinders, which (according to M. Hoefer) is still called, after her, a *Bain-Marie.*

We cannot assign to the Greek MSS. in the Bibliothèque Nationale the antiquity which M. Hoefer and others so readily accept; and we must still hold to our opinion that they and all other known Greek MSS. on alchemy are the production of later centuries, and are probably the work of Greek monks. In the first place, who was Zozimus? Was it Zozimus the Anti-pope, who succeeded Innocent I., or Zozimus the Sophist of Alexandria, or Zozimus the historian? No one can tell. It cannot be pretended that any of the Paris MSS. are in the actual writing of Zozimus. One of them is entitled "Zozimus the Panapolite, on the Chemical Art, to his sister Theosebia;" but, according to the "Biographie Universelle," it was Zozimus of Alexandria who dedicated books to his sister Theosebia, and he lived in the third century B.C., while Zozimus of Panapolis lived in the fourth century A.D. Here, then, we have a discrepancy of 700 years, and a clear confounding of Zozimus of Alexandria with his namesake of Panapolis. Suidas attributes chemical works to the former, but we must remember that the word χημεία does not occur before the eleventh century, A.D. The director of the Bibliothèque Nationale,[1] in a recent letter for which we

---

[1] This Library has so often changed its name of late, that we think it necessary to mention that we mean the Library in the Rue Richelieu, which is called by old writers the *Bibliothèque du Roi,* sometimes the *Bibliothèque Royale,* lately the *Bibliothèque Impériale,* still more lately the *Bibliothèque Communale* now the *Bibliothèque Nationale.* Juncker, in his *Conspectus Chemiæ,* in speaking of various writers on alchemy,

have to thank him, writes as follows :—"La Bibliothèque Nationale ne renferme aucun manuscrit grec de Zosime de Panapolis qui puisse attribué à une époque antérieure au XIII Siècle. Le plus ancien de ceux qu'elle possède ne remonte pas plus loin que cette date." Everything tends to prove that the MSS. were not only written, but composed, at a period posterior to the fifth century. The fanciful titles of some of them show us that their authors adopted any name they pleased ; thus we have " the Epistle of Isis, queen of Egypt, and wife of Osiris, on the Sacred Art, addressed to her son Horus," in which we find a solemn oath dictated to Isis by the angel Amnaël, who swears by Mercury and Anubis, by Tartarus, the Furies, and Cerberus, and by the dragon Kerkouroboros. The whole thing is plainly a blending of Eastern and Western thought : personages of Egyptian, Greek, and Roman mythology, with angels of the Talmud and genii of Arabic lore. We are glad to find that M. Hoefer breaks freely away from the too confident Olaus Borrichius, as to the authenticity of Hermes Trismegistus. He admits that the books which bear his name are spurious, and concludes that their author " vivait probablement à l'époque critique du Christianisme triomphant et du paganisme à l'agonie." But if we take this as the time of Constantine the Great, we must venture to attach a later date to these writings.

We recently had an opportunity of examining the MS. in the Bibliothèque Nationale, attributed to

cites "Zozimus Panapolites celeberrimus et magni cognomen adeptus, cujus varia scripta exstant in Bibliotheca Regia Parisiensi."

Zozimus and to the fifth century; a MS. which, from
its frequent mention in both ancient and modern
works on the history of chemistry, possesses special
interest. It is entitled "Zozimus on Chemical Instruments and Furnaces, and on the Holy Water"
(Ζωσίμου περὶ ὀρχάνων καὶ καμίνων καὶ περὶ τοῦ θείου
ὕδατος), and it is a well-preserved MS. of the thirteenth century, written on vellum. The few drawings
which it contains are asserted to have been taken by
the author from a temple at Memphis. The Alembic
(b in the accompanying woodcut, Fig. 8) is copied

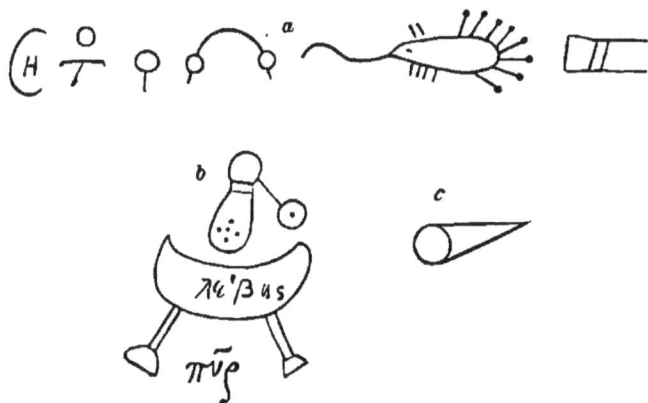

FIG. 8.—An Alembic, and Symbols from Greek MSS. on Alchemy.

from this MS., in which also the line of symbols (a)
is found. These symbols occurred in almost every
Greek MS. on alchemy which we examined; we could
find no clue to the curious porcupine-like animal.
The symbol c is clearly of astronomical origin, and is
not often met with in later works. The MSS. are for
the most part devoid of figures, and not so full of
symbols as later alchemical treatises.

We have endeavoured to prove (a) that no reliable date can be assigned to existing Greek MSS. on alchemy, and (β) that the accepted date is too early. Even if we could prove that a man named Zozimus, living in the fourth century, wrote treatises on alchemy, we could not use the existing MSS. for any exact purpose connected with the history of science with safety; for, since we have no such MS. earlier than the tenth or eleventh centuries, it would be quite impossible to determine whether additions had been made during transcription. The facts are simply these:—There exist in various parts of the world Greek MSS. on alchemy, none of which are older than the tenth century. Many of these bear the names of mythical personages of Egyptian mythology, some of ancient Greek philosophers, some of people who are supposed to have lived in the fourth or fifth century, A.D. When we remember that no ancient writer makes mention of alchemy or chemistry, that the word χημεία is first used in the eleventh century, and when we further bear in mind the condition of the intellectual world in the fourth and fifth centuries, we think we may well admit that further evidence is necessary before we can assert that alchemy arose in the fourth century. Indeed we are of opinion that, in spite of all that has been written on the subject, there is no good evidence to prove that alchemy and chemistry did not originate in Arabia not long prior to the eighth century, A.D.

CHAPTER VII.

Latin and English MSS. on Alchemy—Sources from which the earlier Alchemists acquired knowledge—Arabic learning during the Middle Ages—Geber.

IN the last chapter we discussed the Greek MSS. on alchemy, and endeavoured to show, that, owing to the uncertainty of their age and the obscurity of their authorship, they are less important components of the early history of chemistry than some writers have laboured to prove them.

There exist also many MSS. in Arabic and Persian on alchemy, but in all probability few of them are earlier than the 8th century. The Library of El Escorial is undoubtedly more rich in such MSS. than any existing library; but from the imperfect manner in which its treasures are catalogued, we are unable even to give a list of the more important of these treatises. The British Museum contains several Arabic MSS. on alchemy, written about the 12th century. Such of these as we have seen are devoid of drawings, and apparently also of symbols.

Early MSS. on alchemy in Latin exist in all large libraries. They contain various recipes for making the philosopher's stone, "secrets of art," copies of the inscription of the Smaragdine table, with the interpretation thereof, and an infinite amount of unintelligible nonsense. They differ in no respect from the later printed treatises on alchemy, which we shall presently discuss in detail. The matter of most of the MSS. is to be found in printed works compiled by alchemists of the fifteenth and sixteenth centuries.

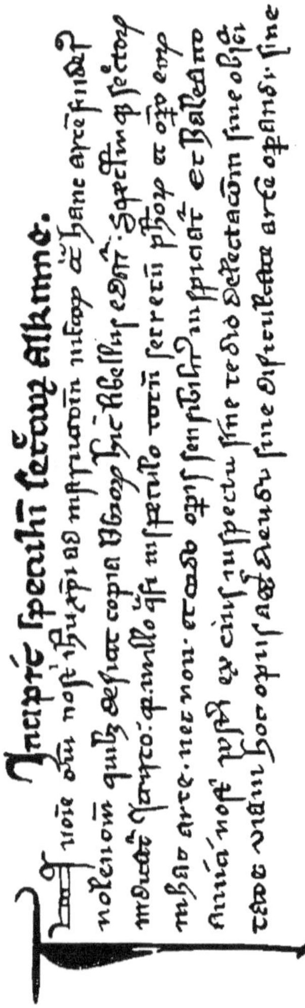

One of the oldest alchemical MSS. in the British Museum is a transcript of the *Speculum Secretorum* of Roger Bacon, who died in 1284. It is in the Sloane Collection, and was written towards the end of the 13th century, say between 1290 and 1300. There is no autograph MS. of Roger Bacon either in the British Museum or in the Record Office; the MS. in question was copied by an unknown

FIG. 9.—Alchemical MS. of the thirteenth century.—British Museum.

man. The woodcut (Fig. 9) represents a few lines of the commencement of the MS., which reads as follows :—" Incipit speculum secretorum alkimiæ. In nomine Domini Nostri Jesu Christi ad instructionem multorum circa hanc artem studere volentium, quibus deficit copia librorum, hic libellus edatur, speculumque secretorum indicatur, idcirco quia in illo, quasi in speculo, totum secretum philosophorum et operatio eorum in hac arte, nec non et ordo operis, sensibiliter inspiciatur. Et habeant amici nostri posteri ex ejus inspectu sine tedio delectationem, sine obscuritate viam hoc opus aggrediendi, sine difficultate artem operandi." The translation is as follows :—" In the name of our Lord Jesus Christ, for the information of the many who wish to devote themselves to the study of this art, and who lack a supply of books, this small manual is published, and is entitled the ' Mirror of Secrets,' seeing that in it, as in a mirror, the whole secret of philosophers and their working in this art—nay more, the process of their work—may be visibly discerned. And may our friendly descendants obtain from the perusal of it unwearied delight, a clear path for taking his work in hand, and a mode of operation unhampered by obstacles."

Among the earlier English MSS. on alchemy in the British Museum is one which, the Preface informs us, was done "at the instance and prayer of a poure creature, and to the helping of man, I, Malmedis, being at greete uneased in prisone, have thees forseide bokes hidre to itake a hand, and so I shal fynnysshe hit, to God be the laude and preisyng."

The following woodcut (Fig 10) represents a portion of this MS. relating to mercury[1] :—

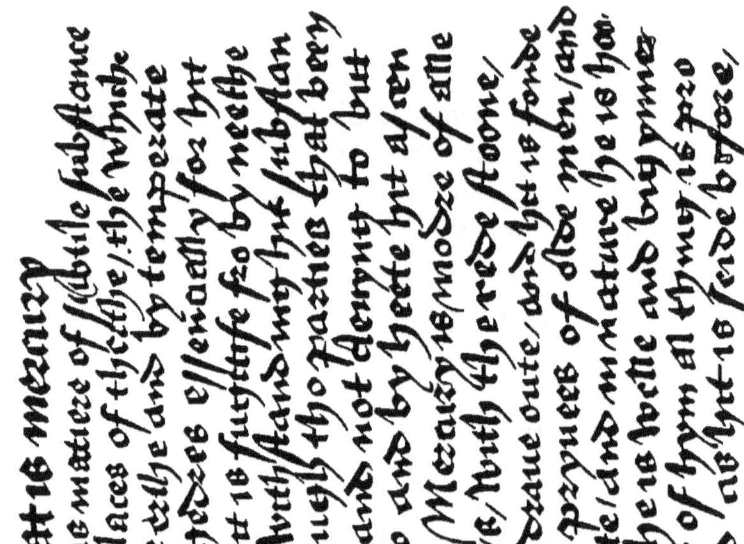

FIG. 10.—English MS. on Alchemy.—Fifteenth century.

[1] We must express our great indebtedness to Mr. Maunde Thompson, of the British Museum, for allowing us ready access to the MSS. department.

It will be noted that mercury, together with sulphur, and the "rede stoone," is designated the producer of all metals ; we also observe an allusion to the Aristotelian theory of the elements (of which an account has been given in the second chapter) in the assertion that mercury is "hotte and moyste." This MS. is in the Sloane collection, and is well preserved, and written on vellum.

Let us now turn our attention to the dogmas of the alchemists and early chemists, as set forth in the numberless printed books on the subject.

We must bear in mind at the outset that chemistry and alchemy—understanding by the former, legitimate inquiry into the nature of different kinds of matter, and by the latter, the futile attempts to make gold— existed side by side in the same age, often in the same person. We cannot agree with M. Hoefer when he says "La chimie du moyen âge, c'est l'alchimie," because some of the early chemists were not alchemists, and the crude processes of the one often led to the exact processes of the other. Lord Bacon, in the *De Augmentis Scientiarum*, has some very pertinent remarks regarding alchemy :—"Credulity in arts and opinions," he remarks, "is likewise of two kinds, viz., when men give too much belief to arts themselves, or to certain authors in any art. The sciences that sway the imagination more than the reason are principally three, viz., Astrology, Natural Magic, and Alchemy. . . . Alchemy may be compared to the man who told his sons that he had left them gold, buried somewhere in his vineyard ; where they by digging found no gold, but by turning up

the mould about the roots of the vines, procured a
plentiful vintage.   So the search and endeavours to
make gold have brought many useful inventions and
instructive experiments to light."

The heritage which the alchemists and early chemists
received from the ancients was by no means insig-
nificant; for they possessed all the experience accu-
mulated by the ancients in the various arts and
processes which we have before described; and of
theoretical matter they possessed, adopted, and
prized, the theory of the transmutation of the elements proposed by Aristotle.   Of works bearing
upon the history of matter, they had the writings
of Aristotle, Dioscorides, Lucretius, Archimedes,
Hero, Vitruvius, and Pliny.   Few books are quoted
more often in alchemic treatises than the "Natural
History" of Pliny; and we sometimes find an almost
verbatim transcript of certain portions of this work.
The alchemists can therefore scarcely be said to have
created a science, for the science of their day is
linked with that of the ancients.

When ancient learning had almost died out, and
Europe was, intellectually, in a state of complete
darkness, the Arabians were the most cultivated
people in the world.   It is to Arabia that we must
look for the origin of several sciences which we are
wont to attribute to other nations.   The Arabians in-
stituted universities, observatories, public libraries, and
museums; they collected together all the remains of
ancient learning, and through their medium the greater
number of Greek and Latin authors which were read
during the Middle Ages were known to Europe.

In the eighth century the Arabs had full possession
of Spain, and at a somewhat later date this country
possessed the most famous universities in Europe.
The Arabs, in propagating their new religion, pro-
pagated also the remains of ancient culture, which
had already been introduced into Persia and Syria
by the Nestorians, who had founded a school of great
reputation at Odessa. Again, when Justinian closed
the schools of Athens and Alexandria, many of the
professors fled to Persia and Arabia, and formed new
centres of learning. The works of many authors,
including Aristotle, Dioscorides, and Pliny, were soon
translated into Arabic and Persian, and became widely
diffused. " Ce fut," remarks M. Figuier, " ainsi que
de l'Inde jusqu'à l'Espagne, des rivages du Tigre
jusqu'à ceux du Guadalquivir, les livres de science se
propagèrent parmi des peuples qui avait déjà une
littérature, un philosophie religieuse, et qui n'étaient
point dépourvus d'imagination."

In the eighth century the University of Bagdad
was founded by the Caliph Al-Mansor, and in the
following century it attained a pre-eminent position.
A large medical school was connected with it, also
hospitals and laboratories. The Caliph Al-Mamoum
erected an observatory in Bagdad, and an attempt
was made to measure an arc of the meridian. It is
said that at one time the University of Bagdad possessed more than six thousand students. In it the
sciences found a home, and every scrap of ancient
learning was eagerly collected and often extended.
When the Arabic empire was broken up by internal dissensions into a number of small states, the

University of Bagdad, losing the powerful patronage
of the Caliphs, fell into decay, and soon ceased to
be known. A somewhat celebrated school arose in
Cairo in the tenth century, but we possess but few
particulars concerning it.

We soon hear of Spain as a focus of learning. In
the tenth century this was the most flourishing
country in Europe, both intellectually and otherwise.
The University of Cordova possessed great celebrity,
and students flocked to it from all parts of the world.
It contained a library of between 200,000 and 300,000
volumes, an unusually large collection of books prior
to the invention of printing. The Arabians were
great mathematicians and astronomers. Lalande
places Mohammed-ben-Giaber (better known as Alba-
tegnius) among the twenty greatest astronomers who
have ever lived. Again, Alhazen wrote a treatise on
optics in the eleventh century, and there were many
treatises on botany and medicine. The Arabs made
but little advance in anatomy however, because they
were forbidden by the Koran to mutilate the human
body.

After the above remarks it is almost needless to
say that we must look to Arabia for the earliest
treatises on alchemy and chemistry. Indeed the
Arabians cultivated the latter science with success,
and the first work on the subject with which we are
acquainted was written by Yeber-Abou-Moussah-
Djafer al-Sofi, whom we call Geber, an Arab of the
eighth century. There had, no doubt, been writers
on chemistry before his time, but probably not long
before. We have endeavoured to prove in the pre-

ceding chapter that the Greek MSS. on the "sacred art" are not trustworthy evidences of the early origin of the science ; and we cannot tell from what source Geber acquired any of his knowledge. He alludes to no one by name, but we know that the Arabians collected knowledge from every source—Egyptian, Indian, Persian, Greek, and Roman. It is thought by some that Geber acquired some of his notions of chemistry from Egypt.

Several MSS. purporting to contain the writings of Geber exist in various libraries in Europe ; these were translated into Latin as early as the year 1529, and into English in 1678. We have reason to believe that the Latin translation was faithfully done, if the Arabic text be not corrupt. The work consists of four treatises :—(a) Of the search for Perfection, (β) Of the Sum of Perfection, (γ) Of the Invention of Verity, and (δ) Of Furnaces.

Geber was acquainted with the seven metals known to the ancients, and he regarded gold, silver, copper, iron, tin, and lead, as compounds of mercury with sulphur in different proportions. Gold and silver are the most perfect metals, and are composed of the purest mercury and sulphur ; the other metals consist of less pure mercury and sulphur, but may be converted into gold and silver by purifying their constituents, and uniting them in different proportions. He describes various chemical substances, among others the following. The carbonates of potash and soda were known to Geber, and were procured from the ashes of plants. Caustic soda was procured from the carbonate by heating its solution with quicklime,

G

as in the present day. Common salt was purified by
ignition, solution, and filtration, and the solution was
afterwards evaporated, and the salt crystallised out.
Nitrate of potash, or saltpetre, and chloride of am-
monium, or sal ammoniac, were apparently common
in Geber's time ; as also were alum, borax, and green

FIG. 11.

copperas, or protosulphate of iron. Geber procured
nitric acid by distilling copperas, saltpetre, and alum,
and, he used the acid for dissolving silver, and when
mixed with sal ammoniac, for dissolving gold. He
obtained nitrate of silver in the form of crystals, and

noticed their fusibility. Various compounds of mercury are described, among others corrosive sublimate
or chloride of mercury, cinnabar or sulphide of mercury, and the red oxide of mercury, in which, nearly
ten centuries later, oxygen gas was discovered by
Dr. Priestley. Geber also obtained sulphuric acid by
distilling alum. He appears to have been acquainted
with the various processes of distillation, sublimation,
calcination, filtration, and many others ; indeed, with
almost all the processes practised by his successors
during the succeeding eight or nine centuries.

It is probable that some of the processes described
by Geber were worked out in the medical schools of
Arabia, and were known shortly before his time ; yet
he was himself a patient worker, and often inter-
sperses his descriptions of substances and processes
with remarks on the method of experimenting, and
the mode of thought most suitable for the studies
which he describes. He has often been called the
" Founder of Chemistry ; " at least his works are the
earliest with which we are acquainted, and he was
venerated as Master alike by the alchemists and
chemists of the Middle Ages.

Geber appears to have been acquainted with many
chemical appliances. In the earliest translations of
his works we find figures of various furnaces and
forms of distillatory apparatus ; one of them, not
unlike a still now in use, is represented above
(Fig 11). The greater number of vessels described
and figured by Baptista Porta in his treatise " De
Distillationibus," published in 1609, are to be found
in the first Latin translations of the works of Geber.

CHAPTER VIII.

Avicenna—Albertus Magnus—S. Thomas Aquinas—Roger Bacon—Raymond Lulli—Arnoldus de Villâ Novâ—George Ripley—Basil Valentine.

THE Schools and Colleges of Arabia soon gave evidence of their value by the development of several considerable geniuses, whose works formed the textbooks of Europe during a portion of the Middle Ages. Prominent amongst these learned Arabians was Ali-ben-Sina, or Avicenna, who was born in 980, in the neighbourhood of Shiraz. His abilities were considerable, and no pains were spared in his education ; as a boy he read the Almagestum of Ptolemy, the Geometry of Euclid, and the Philosophy of Aristotle, and later in life he studied medicine with great success. We are told indeed that at the age of sixteen he was an eminent physician, and that at eighteen he cured a caliph of some grave disorder, and was hence promoted to great honour.

Avicenna is best known by his celebrated " Canons," which were translated at an early date into Latin, and often printed under the title of " Canones Medicinæ." This work has been translated into the languages of all civilised countries, and for no less than six centuries was the standard medical treatise of the world.

Avicenna also wrote on alchemy and on chemistry. If the works attributed to him are genuine, he appears to have adopted the Aristotelian theory of the four mutually convertible elements. He speaks of air as the aliment of fire, and of the metals as compounds of a humid substance and an earthy substance. This last idea evidently arose from the observation of the calcination of metals. It was well known that if certain metals, such as lead and tin, are heated for a length of time in the air, they are converted into a powdery substance or calx, and it was long before it was proved that this calx is not the metal from which one of its constituents has been expelled by fire ; but, on the other hand, the metal combined with another substance. Avicenna divides all minerals into four classes : viz. (1) Infusible minerals ; (2) Minerals which are fusible and malleable, that is, metals ; (3) Sulphurous minerals ; and (4) Salts. He noticed that mercury can, by heat, be caused to unite with sulphur and produce a solid body, having different properties from its constituents.

Avicenna was largely indebted for his knowledge to Alfarabi and to Rhazes. The latter wrote on medicine, and was one of the first to introduce substances formed artificially by chemical means into medicine.

Turning now our attention to European alchemists, we meet at the outset with the name of Albertus Magnus (b. 1193, d. 1282), who became Bishop of

Ratisbon in 1259. Various works on alchemy are attri
buted to him ; he wrote on the Philosopher's Stone,
on the origin of metals, and on minerals ; and he has
described at some length various chemical operations,
such as sublimation and distillation, and various forms
of apparatus, such as aludels, alembics, and water-
baths. He followed Geber in the belief that metals
are composed of sulphur and mercury, and that
different metals are produced by different combin-
ations, and to some extent by the variations in
the purity, of these substances. Albertus Magnus
employs the term *affinity* (*affinitas*) to designate the
cause of the combination of sulphur with silver and
other metals; in this precise sense, applied to all
cases of chemical combination, the term is used in the
present day. He also speaks of sulphate of iron as
vitriol, a name which it long retained. He describes
the preparation of nitric acid, its principal effects
upon certain metals, and its utility for separating silver
from gold, inasmuch as it will dissolve the former and
not the latter. Cinnabar, or sulphide of mercury,
had long been known and used as a source of mercury ;
Albertus proved that it consists of sulphur and
mercury by preparing it artificially, by subliming
sulphur with mercury.

Albertus was not alone learned in alchemy ; he
was a profound theologian, a scholar, an astronomer,
a physician, and some said an adept in magic and
necromancy. He embodied his wisdom in twenty-one
folios, which were published in a collected form in
1651. M. Lenglet Dufresnoy, in his " Histoire de la
Philosophie Hermétique," has mentioned several

magical operations gravely attributed to Albertus
Magnus by various writers. The most noticeable piece
of magic was the sudden transformation of a winter's
day into glowing summer :—" Horridam hyemem,"
says Trithemius, " in florigeram fructiferamque vertit."
It is said that once during a very severe winter, he
invited Count William of Holland, when he was passing through Cologne, to a feast. The Count, on his
arrival with a considerable retinue, was surprised to
find the feast spread in the garden, in which there
was a depth of several feet of snow ; and this treatment so angered him that he remounted his horse and
prepared at once to leave his inhospitable host.

> " Then the monk falling on his knees besought
> The Count to sit one moment at the board.
> He having done so, a most wondrous change
> Passed on the instant over all around.
> The dark clouds floated off and left a sky
> Intensely blue, an air exceeding clear ;
> The sun shone brightly, and the warm south wind
> Laved their pale cheeks and warmed them into life.
> They sit on greenest grass, the snow is gone,
> Sweet flowers bloom beneath their very feet,
> Ripe peaches blush upon the garden wall,
> And orange blossoms scent the humid air.
> A swarm of insect life on droning wing
> Is floating up above them in the breeze.
> The voice of birds is heard ; the cooing dove
> Speaks softly to her mate ; the nightingale
> Trills a sweet lay, half hidden in the leaves.
> All nature is most joyous in her garb
> Of brightest summer day, and all things seem
> To glory in the flood of warmth and light."

Upon this, the Count expressed considerable astonish-
ment, as, although he had heard a good deal of the
magical powers of his host, he was quite unprepared

to find him capable of changing the seasons. As soon as the feast was ended, Albertus repeated a magical formula—

> " Now snow obscures the air, the flowers fade,
> The trees are torn by pitiless strong winds,
> And weep their shrivelled fruit upon the earth:
> All sound of life is gone, a roar of elements
> Succeeds the plaintive quavering of the leaves.
> The birds fall dead to earth, and the dark air
> Betokens fearful tempests yet to come."

So the Count and his retinue rush off into the house to warm themselves, and thus ends the feast of Albertus Magnus. Some will have it that the story alludes to a winter garden, which had been devised by Albertus for the preservation of rare plants, and which was unknown at that time. Middle Age books on science abound with such stories, and the belief in them was almost universal, as it well might be in an age in which the power of witches and wizards was acknowledged, and the raising of the dead was an admitted possibility. Brücker (*Institutiones Historiæ Philosophicæ*) says :—" Quæ enim de ejus convivio magico narrantur, merito inter inficeti seculi fabulas referuntur, quæ ex ignorantia rerum naturalium eo tempore crassissima et Alberti mirabili rerum physicarum cognitione prodierunt."

In the church of S. Andreas in Cologne they show to this day the shrine and relics of Albertus—the accomplished churchman, scholar, magician and alchemist, of whom Trithemius says, " Magnus in Magia Naturali, major in Philosophia, maximus in Theologia."

Albertus had for his pupil the "angelic doctor,"
S. Thomas Aquinas (b. 1225, d. 1274), who was
a great alchemist, and who wrote a treatise called
"The most secret Treasure of Alchemy," together
with some other works on the subject, which are
equally obscure and unintelligible. He wrote also on
the artificial preparation of gems, by fusing glass with
certain substances, like oxide of copper, to com-
municate different colours ; he mentions that if copper
be heated with white arsenic, the former becomes
white, something like silver. According to some,
S. Thomas Aquinas was the first to employ the term
amalgam, to designate a compound of any metal with
mercury. S. Thomas Aquinas was, like his master,
a magician. We are told that between them they
constructed a brazen statue, which Albertus animated
with his *elixir vitæ*. It was useful as a domestic ser·
vant, but very talkative and noisy ; nor could they
cure it of this propensity. It happened one day that
S. Thomas, who was a mathematician, was deeply
engaged in a problem, but was continually interrupted
by the talking statue ; at length in a rage he seized
a hammer and smashed it to atoms, to the great
regret of his master.

Our great countryman Roger Bacon (b. 1214) also
suffered from a charge of magic, and during his residence in Oxford was severely persecuted in con-
sequence. He replied to the charges made against
him by the admirable treatise " De nullitate Magiæ,"
and in it clearly showed that what his contemporaries
mistook for the work of spirits, was in good sooth due
to the ordinary operations of Nature. In this work he

speaks of gunpowder, although somewhat obscurely. "Mix," says he, "together saltpetre, *luru vopo vir con utrict*, and you can make thunder and lightning, if you know the method of mixing them." Elsewhere he says, " a small quantity of matter properly manufactured, and not larger than one's thumb, may be made to produce a horrible noise and sudden flash of light." The third constituent of gunpowder is designated under the anagram *luru vopo vir con utriet*, for it was dangerous in those days to speak too plainly ; indeed Bacon tells us that he adopted an obscure style both on account of the example of other writers, and of propriety, and also on account of the dangers of plain speaking. According to some writers, the following passage is to be found in Bacon's writings :—" Sed tamen salis petræ, *luru mone cap ubre*, et sulphuris, et sic facies tonitrum si scias artificium." Thus the saltpetre and the sulphur are directly designated, while the anagram *luru mone cap ubre* is convertible into *carbonum pulvere*, the remaining constituent powdered charcoal. It is improbable that Roger Bacon invented gunpowder, although he was the first to know of its properties in England ; he probably procured the knowledge from an Arabic source. Gunpowder was first used by the English at the battle of Crecy in 1346, sixty-one years after the death of Bacon ; at this time it was apparently unknown to other European nations.

Roger Bacon is believed to have been far in advance of his times in all matters of science. To him has been attributed the invention of the telescope and *Camera obscura*, and several discoveries of a later date. The evidence is less conclusive than one could wish, but enough remains in his writings to prove that he was a very learned man and profound thinker. His "Opus Majus" clearly proves that he fully recognised the value of the experimental method, and of the inductive philosophy afterwards so ably advocated by his namesake Francis Bacon. Roger Bacon wrote largely on alchemy. Many of the alchemical MSS. in the British Museum are transcripts of portions of his works, among the more celebrated of which we may mention the "Medulla Alchymiæ," "Secretum Secretorum," and "Speculum Secretorum." He collected together the principal alchemical facts of his predecessors, and appears in many matters to have closely followed Geber. Bacon describes the distillation of organic substances, and alludes to the inflammability of the evolved gases. He proved that air is the food of fire by burning a lamp in a closed vessel.

Raymond Lulli (b. 1235) is by some asserted to have been a pupil of Roger Bacon. He was a voluminous writer on alchemy, his most celebrated treatise being his "Ultimum Testamentum." He also wrote on transmutation, on the Philosopher's Stone, and on magic. Lulli does not appear to have added to the chemical knowledge of his predecessors ; he followed Geber closely, and was well acquainted with the processes and compounds which he describes. He describes alcohol under the names of *aqua vitæ ardens*, and *argentum vivum vegetabile*, and was in the habit of rendering it anhydrous by allowing it to stand in contact with dry carbonate of potassium. He was also acquainted with ammonia.

Whatever Lulli's knowledge may have been, he obtained great reputation as a successful alchemist. He
asserts in his " Ultimum Testamentum " that he converted fifty thousand pounds weight of base metals
into gold. He is said to have been employed by one
of the Edwards to make gold, and to have furnished
His Majesty with six millions of money. Dickenson
tells us that Lulli had a laboratory in Westminster
Abbey, in which, after his departure, a quantity of
gold dust was found.

Of the general tone and character of alchemical
writings we shall speak more fully in the next chapter.
Of the professors of the art little more need be said ;
a long list of names might be given, but it would be
found that they did little to develop what afterwards
became the science of chemistry. Let us glance at
the work of a few of the remaining alchemists. Arnoldus de Villâ Novâ (b. 1240) was a great alchemist
and physician, and the author of many works on the
subject. His " Rosarius Philosophorum " purported
to contain a key to all alchemical operations. He
followed Geber closely. He considered a solution of
gold the most perfect medicine, and we usually find
that such solution was recommended by alchemists
as a necessary constituent of the *elixir vitæ*, and
essential for the work of transmutation. In Fig. 12
the solution of gold in the flask is represented by the
sun emitting rays. The simple disc of the sun is
the more common symbol for gold.

Arnoldus also distilled various oils and essences.
He contended that sulphur, arsenic, mercury, and sal
ammoniac—all volatile bodies be it noted—are the

souls of metals, and are given off during calcination.
He also affirmed that silver is intermediate between
mercury and other metals, just as the soul is intermediate between the spirit and the body.   Arnoldus
is said to have had for his pupil Pope John XXII., an
accomplished alchemist. who left at his death eighteen

FIG. 12.—An alchemist hermetically sealing a flask containing a solut'on of gold.

millions of florins, which the alchemists fondly cite as
a proof of the possibility of transmutation.

Our countryman, George Ripley, Canon of Brid-
lington in Yorkshire (b. about 1460), wrote a poem on
alchemy, and passed for a successful disciple of the
art, but we cannot point to a new fact which he
elucidated.   He divided all chemical operations into
twelve processes—calcination, dissolution, separation,

conjunction, putrefaction, congelation, cibation, sublimation, fermentation, exaltation, multiplication, and
projection. Several MS. copies of his poem exist in
the British Museum, bound up with copies of the

FIG. 13.—Alchemical representation of processes.

works of Roger Bacon and earlier writers. Here is a
specimen of his rugged rhymes:—

> The fyrst chapter shall be of naturall *Calcination* ;
> The second of *Dyssolution*, secret and phylosophycall;
> The third of our elementall *Separation* ;
> The fourth of *Conjunction* matrimoniall ;
> The fyfth of *Putrefaction* then followe shall :
> Of *Congelation Alhyficative* shall be the sixt,
> Then of *Cybation*, the seaventh shall follow next.

One of the most celebrated of the alchemists was
Basil Valentine, who was born at Erfurt in 1394

According to Olaus Borrichious, his works were accidentally discovered in the wall of a church at Erfurt many years after his death. A thunderbolt struck the church and exposed to view the long-lost alchemistical treasures. Basil Valentine was the author of many treatises, the most important being his "Currus Triumphalis Antimonii," in which he discusses the properties of antimony and of some of its compounds. He regarded the metals as compounds of salt, sulphur, and mercury; and he was acquainted with many metallic compounds, among others nitrate of mercury, sulphide of arsenic, red oxide of mercury, chloride of iron, sulphate of iron, fulminating gold, carbonate of lead, acetate of lead, and the oxides of lead. He was aware that iron precipitates copper from solution, and that solution of potash precipitates iron from solution. He was well acquainted with the preparation of nitric and sulphuric acids, and used them for various purposes of dissolution. In order to obtain nitric acid he distilled powdered earthenware with nitre, or equal weights of nitre and green vitriol, or nitre with finely powdered flints. He obtained fuming sulphuric acid by distilling green vitriol, after the manner still practised at Nordhausen and elsewhere. Basil Valentine wrote very obscurely and was fond of symbolical designs. Figures 13 and 14 are taken from his works, and represent various processes imperfectly described. Thus the lion in Fig. 13 would represent a solution of a metal, the serpent another solution, or perhaps the serpent a metal, and the lion devouring it a solvent; the sun and moon are watching the

operation, and the symbol of mercury appears between
two roses.   Fig. 14 represents some operation which
is thus described by the principal figure :—" I am an
old, infirm, debilitated man, my soul and spirit
(represented by the two boy-headed birds above
his head) leave me, and I assimulate the black

FIG. 14.—Alchemical representation of processes.

crow.   In my body are found salt, sulphur, and mercury." This may possibly refer to the solution of gold
in aqua regia ; it loses its metallic nature, its solidity
and lustre, and assimulates the acid ; but one may
conjecture in vain concerning the enigmatical devices
in which some of the alchemists took so much delight,

and which they often employed, like Roger Bacon's anagram, to conceal the full significance of their operations or discoveries.

The following extract, which treats of the generation of metals, will show the style of Basil Valentine's writing :—

" Therefore think most diligently about this ; often bear in mind, observe, and comprehend that all minerals and metals together in the same time, and after the same fashion, and of one and the same principal matter are produced and generated. That matter is no other than a mere vapour, which is extracted from the elementary earth by the superior stars or by a sidereal hot infusion, with an airy sulphureous property, descending upon inferiors, so acts and operates as in those metals and minerals is implanted spiritually and invisibly a certain power and virtue, which fume afterwards resolves in the earth into a certain water from which mineral water all metals are thenceforth generated and ripened to their perfection, and thence proceeds this or that metal or mineral according as one of the three principles acquires dominion, and they have much or little of sulphur and salt, or an unequal mixture of them ; whence some metals are fixed, that is constant or stable ; some volatile and easily mutable, as is seen in gold, silver, copper, iron, lead, and tin."

Now this is by no means the most obscure piece of alchemical writing with which we shall come in contact.

H

CHAPTER IX.

*General Character of Alchemy and the Alchemists—The " Pre-
tiosa Margarita Novella"—An Alchemistical Allegory—
Alchemical Symbols—Paracelsus—Libavius.*

WHAT manner of men were the alchemists? How
did they preserve, cultivate, and transmit the wonderful delusions of their creed? We have endeavoured
in a former chapter to show that the idea of transmutation arose from the old Greek idea of the conversion of one element into another; and the belief
in the possibility of transmutation once admitted, the
pursuit of the alchemist would naturally follow in a
mystical and credulous age. As to the men themselves, their character was twofold; for there was
your alchemist proper, your true enthusiast, your
ardent, persevering worker, who believed heart and
soul that gold could be made, and that by long search
or close study of the works of his predecessors, he
could find the Philosopher's Stone; and there was
your knavish alchemist, a man who had wits enough
to perceive that the search was futile, and impudence
enough to dupe more credulous people than himself
and wheedle their fortunes out of them on pretence

of returning it tenfold in the shape of a recipe for converting lead into gold. These last we may dismiss at once. They abounded during the Middle Ages, and found easy dupes, whom they deceived by the most shallow tricks, as by placing a piece of gold in the crucible •of transmutation together with volatile substances, and after many processes and much heating, they would show the little button of metal which had all along been present.

Of the true alchemist we have many pictures. The alchemist, the astrologer, the mystic, the wizard, were men of the same stamp. They often practised the same arts side by side. The same habit and attitude of thought belonged to one and to all, and became all equally well. Take the dreamy, maudlin, semi-maniacal Althotas, who has been described so well by Dumas :—"An old man, with grey eyes, a hooked nose, and trembling but busy hands. He was half-buried in a great chair, and turned with his right hand the leaves of a parchment manuscript." Note also his intense abstraction, his forgetfulness of the hour, the day, the year, the age, the country ; his absolute and intense selfishness and absorption, the concentration of the whole powers of his soul upon his one object. Or let us look at Victor Hugo's Archidiacre de St. Josas, in his search for the unseen, the unknown, and the altogether uncanny ; the bitterness of his soul, his passionate musings, his conjurations and invocations in an unknown tongue ; his own self, that wonderful mixture of theologian, scholar mystic, perhaps not much unlike the divine S. Thomas Aquinas himself. Listen to his musings :

" Yes, so Manon said, and Zoroaster taught :—the sun is born of fire, the moon of the sun; fire is the soul of the universe; its elementary particles are diffused and in constant flow throughout the world, by an infinite number of channels. At the points where these currents cross each other in the heavens they produce light, at their points of intersection they produce gold. Light!—gold! the same thing; fire in its concrete state. . . . . What! this light that bathes my hand is gold? The first the particles dilated according to a certain law, the second the same particles condensed according to another law! . . . For some time, said he, with a bitter smile, I have failed in all my experiments ; one idea possesses me, and scorches my brain like a seal of fire. I have not so much as been able to discover the secret of Cassiodorus, whose lamp burned without wick or oil —a thing simple enough in itself." If we peep into Dom Claude's cell, we are introduced to a typical alchemist's laboratory—a gloomy, dimly-lighted place, full of strange vessels, and furnaces, and melting-pots, spheres, and portions of skeletons hanging from the ceiling ; the floor littered with stone bottles, pans, charcoal, aludels, and alembics, great parchment books covered with hieroglyphics ; the bellows with its motto *Spira, Spera;* the hour-glass, the astrolabe, and over all cobwebs, and dust, and ashes. The walls covered with various aphorisms of the brother-hood ; legends and memorials in many tongues ; passages from the Smaragdine Table of Hermes Trismegistus ; and looming out from all in great capitals, 'ΑΝΑΓΚΗ. Yet once again, look at Faust, as depicted by Rembrandt; or Teniers' unknown alchemist, if you wish for an alchemical interior.

But the hard-working and enthusiastic alchemist did not always follow the ideal of the novelist and artist; he often degenerated into a "dirty soaking fellow," who lost what little learning he ever had by concentrating his mind on the one dominant topic, until it excluded every other idea and aspiration; then the pursuit became all-absorbing, and the disciple of the art a mere drivelling monomaniac.

We will now look at one of the books which were cherished by the alchemists. Here is a little vellum-covered *Aldus:* date 1546. Paracelsus had been dead five years, and Cornelius Agrippa twelve years; Dr. Dee and Oswald Crollius were flourishing; Van Helmont and a host of known alchemists were unborn. Our little volume, full of quaint musings of a bygone age, has outlived them all, and yet it never drank of the *elixir vitæ*, although it pretended to teach others how to make it, and the Philosopher's Stone into the bargain. "Pretiosa Margarita Novella de Thesauro, ac pretiosissima Philosophorum Lapide" is the title; published with the sanction of Paul III., Pontifex Maximus, whose successor, be it remembered, established the "Index Expurgatorius," and might possibly have prohibited this Precious Pearl of alchemy. The title-page tells us that it contains the methods of the "divine art," as given by Arnoldus de Villâ Novâ, Raymond Lulli, Albertus Magnus, Michael Scotus, and others, now first collected together by Janus Lacinius. The vellum cover is well thumbed, and in one place worn through, perhaps by contact with a hot iron on an alchemist's furnace-table, or by much use. There are no MS. notes, but on the title-page is the autograph of Sir E. Koby, or Hoby, and a favourite maxim, the first word of which—*Fato*—is alone legible. The date of the writing is perhaps 1580–90. Some initial letters of the text have been plainly illuminated in red, by a loving hand; they were copied from a Bible transcribed at Lyons in 1326.

Fig. 15.—Allegorical representation of transmutation.

As to the contents, we have firstly an opening address by Janus Lacinius; then certain definitions of form, matter, element, colour, &c.; next, symbolic representations of the generation of the metals, and after this a woodcut representing the transmutation of the elements according to the dogmas of Aristotle.[1] After this we find the whole course of transmutation

[1] See Chapter I. Fig. 1.

set forth pictorially and allegorically, as under. A
king (see Fig. 15), crowned with a diadem, sits on
high, holding a sceptre in his hand. His son, together
with his five servants, beseech him on bended knees
to divide his kingdom between them. To this the
king answers nothing. Whereupon the son, at the
instigation of the servants, kills the king and collects
his blood. He then digs a pit, into which he places
the dead body, but at the same time falls in himself,
and is prevented from getting out by some external
agency. Then the bodies of both father and son

Fig. 16.—Allegorical representation of transmutation.

putrefy in the pit. Afterwards their bones are removed, and divided into nine parts, and an angel is
sent to collect them. The servants now pray that
the king may be restored to them, and an angel

vivifies the bones. Then the king rises from his tomb, having become all spirit, altogether heavenly and powerful to make his servants kings. Finally he gives them each a golden crown, and makes them kings (Fig. 16).

It is difficult to follow this from beginning to end, but there can be no doubt that the king signifies gold, his son, mercury, and his five servants the five remaining metals then known, viz. iron, copper, lead, tin, and silver. They pray to have the kingdom divided amongst them, that is to be converted into gold ; the son kills the father, viz. the mercury forms an amalgam with gold. The other operations allude to various solutions, ignitions, and other chemical processes. The *pit* is a furnace ; *putrefaction* means reaction or mutual alteration of parts. At last the Philosopher's Stone is found ; the gold, after these varied changes becomes able to transmute the other metals into its own substance. At the end some rugged hexameters and pentameters warn the fraudulent, the avaricious, and the sacrilegious man that he is not to put his hands to the work, but to leave it for the wise and the righteous, and the man who is able rightly to know the causes of things.

After this allegory we have some remarks concerning the Treasure, and the Philosopher's Stone, and the Secret of all Secrets, and the Gift of God. This is followed by a number of arguments against alchemy, and of course overwhelming arguments in favour of it. Among those who are quoted as alchemists are Plato, Pythagoras, Anaxagoras, Democritus, Aristotle, Morienus, Empedocles, and then, with a delightful

disregard of age or country, we read, "Abohaby,
Abinceni, Homerus, Ptolemæus, Virgilius, Ovidius."
Then digressions on the difficulties of the art, the
unity of the art, the art natural and divine; a slight
history of the art, in which it is traced back to Adam,
although Enoch and Hermes Trismegistus are mentioned as possible founders. A treatise to prove that
this art is more certain than other sciences; on the
errors of operation ; on the principles of the metals ;
on sulphur ; on the nature of gold and silver; and
many general remarks on all alchemical subjects.
These are the teachings which the "Pretiosa Margarita
Novella" pours at the feet of the wise among mankind,
by the aid of Paulus Manutius, bearing his father's
name of Aldus, and by the grace of the Venetian
Senate.

Many attempts were made by the alchemists to
explain the origin of the metals. Some regarded them
as natural compounds of sulphur and mercury; others
affirmed that the power of the sun acting upon and
within the earth produced them, and that gold was
in truth condensed sunbeams. Many believed that
metals grew like vegetables ; indeed it was customary
to close mines from time to time to allow them to
grow again. Basil Valentine, as we have seen, regarded them as condensations of a " mere vapour into
a certain water," by which latter we suspect he meant
mercury. Perhaps the most absurd account of the
origin of certain things is given by Paracelsus in his
treatise, "De Natura Rerum," in the following words,
which will show also how utterly nonsensical and
unintelligible alchemical language could be, and for

that matter very generally was. "The life of metals,"
he writes, "is a secret fatness; . . . of salts, the spirit
of aquafortis; . . . of pearls, their splendour; . . .
of marcasites and antimony, a tinging metalline
spirit; . . . of arsenics, a mineral and coagulated
poison . . . The life of all men is nothing else but an
astral balsam, a balsamic impression, and a celestial
invisible fire, an included air and a tinging spirit of
salt. I cannot name it more plainly, although it is
set out by many names."

The peculiarly secret and mystical language which
the alchemists adopted was intended to prevent the
vulgar from acquiring the results of their long-con-
tinued labours. Their language purported to be
intelligible to the true adept; but as a rule the alchemists of one age gave various interpretations to one
and the same secret communicated by their predecessors. Long recipes for the preparation of the
Philosopher's Stone exist, which the authors have
generously (as they tell us) given to the world, after
much labour, for the benefit of their fellow-men.
The obscurity of the science was increased by the
multiplication of symbols; the presence of which in
alchemy clearly points to its connection with astrology
and the sister sciences. In time alchemical symbols
multiplied almost as much as astrological symbols.
In an Italian MS. of the early part of the seventeenth
century which we have before us, mercury is represented by 22 distinct symbols, and 33 names, many
of which are of distinctly Arabic origin :—such as
Chaibach, Azach, Jhumech, Caiban. Lead is represented by the symbols in Fig. 17, and in addition to

its ordinary alchemical names, is called Okamar,
Syrades, Malochim, and others.   The designation of
substances as "the green lion," "the flying eagle,"
"the serpent," "the black crow," and so on, also led
to considerable confusion.   Both names and symbols
were used in a somewhat arbitrary fashion.

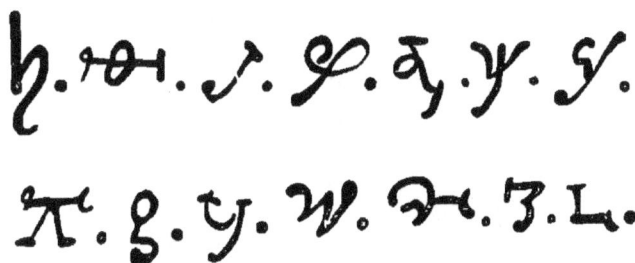

FIG. 17.—Symbols of lead from Italian MS. of the seventeenth century.

It is somewhat strange to think that alchemy should
have once received the serious attention of the legis-
lature in this country.   In 1404 the making of gold
and silver was forbidden by Act of Parliament.   It
was imagined that an alchemist might succeed in his
pursuit, and would then become too powerful for the
State.   Fifty years later Henry VI. granted several
patents to people who thought they had discovered
the Philosopher's Stone; and ultimately a commission
of ten learned men was appointed by the King to
determine if the transmutation of metals into gold
were a possibility.   We must now leave the subject
of alchemy.   Those who desire to study it more
deeply will find a great mass of matter in the "Bibliotheca Chemica Curiosa" of Mangetus; but if they
will take our advice, they will not waste much time

in studying the history and progress of a futile and false art.

With Paracelsus (b. 1493 d. 1541) a somewhat new phase of the science of chemistry appeared. By pointing out the value of chemistry as an adjunct to medicine, he caused a number of persons to turn their

FIG. 18.—Designs from Mangetus (*Bibliotheca Chemica Curiosa*).

attention to the subject, and to endeavour to ascertain the properties of various compounds. Thus he helped to withdraw men from the pursuit of alchemy, by asserting that the knowledge of the composition of bodies, which had necessarily been forwarded by

alchemy, was of importance to the human race, for
the better prevention and curing of their ills. In the
way of discovery or research, Paracelsus did little.
He mentions zinc and bismuth, and associates them
with metallic bodies, and he makes considerable use
of several compounds of mercury, and of sal ammoniac. Paracelsus compares the alchemist of his
day with the physician, and speaks of the former in
the following terms:—"For they are not given to
idleness, nor go in a proud habit, or plush and velvet
garments, often showing their rings upon their fingers,
or wearing swords with silver hilts by their sides, or
fine and gay gloves upon their hands, but diligently
follow their labours, sweating whole days and nights
by their furnaces. They do not spend their time
abroad for recreation, but take delight in their laboratory. They wear leather garments with a pouch,
and an apron wherewith they wipe their hands. They
put their fingers amongst coals, into clay, and filth,
not into gold rings. They are sooty and black like
smiths and colliers, and do not pride themselves upon
clean and beautiful faces."

Among the Paracelsians we find Oswald Crollius,
who mentions chloride of silver under the long-
retained name of *luna cornea*, or horn-silver, from its
peculiar horny appearance and texture after fusion.
He was also acquainted with fulminating gold.

The name of Andrew Libavius (died 1616) deserves
mention, because he sought to free chemistry from
the mazes of alchemy and mysticism in which it was
involved. In this he to some extent succeeded ; and
he appears also to have been a patient worker in the

field of the science which he did so much to promote. He discovered the perchloride of tin, which is even now called *fuming liquor of Libavius;* he also proved that the acid (sulphuric acid) procured by distilling alum and sulphate of iron, is the same as that prepared by burning sulphur with saltpetre. Libavius was great at the making of artificial gems, and was able to imitate almost any precious stone by colouring glass with various metallic oxides.

CHAPTER X.

As in the history of matter we find molecules grouping themselves around a common centre or a common line, thus constituting crystalline bodies, so in the history of sciences and of nations we may often observe well-defined axes, about which the facts of particular epochs congregate. Such axes are to be found in the history of chemistry. At the particular period of which we now write, the facts of the science mainly grouped themselves around theories connected with combustion, which involved as collateral matters conceptions regarding the nature of calcination and of the air.

Combustion was, and still is, the most prominent exhibition of chemical force, with which man ordinarily comes into contact. It is a purely chemical action—the union of dissimilar bodies under the influence of chemical affinity, attended by the evolution

of light and heat. Many attempts were made to
explain its cause. Fire, in common with earth, air,
and water, as we have before seen, was regarded as
an element, till almost within our own memory.
Epicurus regarded heat as a congeries of minute
spherical particles possessing rapid motion, and readily
insinuating themselves into the densest bodies. Fire
was simply an intense form of heat. Cardanus speaks
of flame as *aer accensus*, and of fire as heat immensely
augmented. During the Middle Ages the existence
of two kinds of fire was admitted—the one pure
celestial fire "*subtilis ignis*," "*cœlestis ignis*," the
principle or essence of fire; the other "gross earthly
fire," or "mundane fire." The latter was the *materia*,
the former the *forma*. Celestial fire became mundane
fire when it was associated with combustible bodies,
that is, in ordinary combustion. Seneca tells us that
the Egyptians divided each element into an active
and a passive form; fire became active flame which
burns, and comparatively passive warmth and light.
The elemental nature of fire was not universally admitted during the Middle Ages; thus Francis Bacon
asserts, in the "Novum Organum," that fire is "merely
compounded of the conjunction of light and heat in
any substance," and he defines heat as a rapid motion
of material particles. Athanasius Kircher, in his
ponderous treatise, "Ars Magna Lucis et Umbræ,"
affirms that fire is air which is caused to glow by the
violent collision of bodies, by which means combustible bodies become flame. At an early date it
was observed that fire cannot exist without air; the
experiment of burning a candle in a closed vessel was

well known. Some affirmed that "air is the food of fire," some that "air nourishes fire." The influence of a blast of air upon fire was well recognized ; we have seen that bellows were known at a very early date. When nitre—which for many centuries was one of the most important bodies in chemistry—came to be known, it was soon noticed that it produces intense ignition ; that, in fact, to direct a blast of air upon a red-hot coal, or to throw some nitre upon it, produced the same result, viz. greatly augmented combustion. Hence arose the idea that nitre and the air are in some way connected, for "things which are equal to the same are equal to each other." This association of ideas may seem crude to us now, yet we must remember that nitre produces rapid combustion simply because it contains a great quantity of that constituent of the air, oxygen gas, which ordinarily produces combustion. Thus the old natural philosophers, wandering in the dim twilight of experimental knowledge, were not so far wrong in their supposition. The idea mentioned above was very prevalent two centuries ago: Robert Boyle speaks of the presence of a "volatile nitre" in the air ; Lord Bacon says that nitre contains a "volatile, crude, and windy spirit"; Clark attributed thunder and lightning to the presence of nitre in the air ; Gassendi imagined that minute particles of nitre are diffused throughout the atmosphere. When we heat lead or tin in a current of air, these metals are respectively converted into a powder, or *calx*, and calcination was one of the most important processes in old chemistry. Calcination seemed to be due more or less directly to the

I

air; and metals could also be calcined by heating them with nitre, or with the spirit of nitre—nitric acid; hence arose another bond of connection between nitre and the air; at least, they had something in common. Lemery in his *Cour de Chimie*, published in 1675, affirms that the acid of nitre contains a number of "*corpuscules ignées*" locked up in it, and he defines these latter as "a subtle matter, which having been thrown into a very rapid motion, still retains the power of moving with impetuosity, even when it is enclosed in grosser matter; and when it finds some bodies which by their texture or figure are apt to be put into motion, it drives them about so strongly that, their parts rubbing violently against each other, heat is thereby produced."

Thus recognizing the causes which had led to the association of the air with nitre, at least so far as they are both concerned in the production of combustion, we are prepared to examine Robert Hooke's theory of combustion. The announcement of this theory marks an important epoch in the history of chemistry; it was the first chemical theory worthy of the name, and it gave a far more just and accurate explanation of combustion than the crude and over-belauded theory of Phlogiston, of Beccher and Stahl. Hooke's theory was, moreover, founded upon ex-periment, and although unfortunately he does not describe the experiments, we see at a glance that it could not have been constructed without such means. "This hypothesis," he writes, "I have endeavoured to raise from an infinity of observations and experiments," and all who know Hooke's writings are well aware how good an experimenter he was. The theory was published in 1665 in Hooke's *Micrographia ;* it is there found (Observation 16) buried in a mass of irrelevant matter, and to this cause may, perhaps, to some extent be attributed the fact that it has been so little recognized and known. The theory is stated in twelve Propositions, the principal of which are as follows :—

1. "That the air is the universal dissolvent of all sulphurous bodies."

Sulphur was long regarded as the type of combustible bodies, on account of its ready inflammability ; some even derive the name from *sal, πῦρ,* the salt of fire. By sulphurous bodies, Hooke simply meant combustible bodies, viz. bodies that can burn in a supporter of combustion. By air being the " universal dissolvent," he meant that through the agency of air combustible bodies are caused to become transformed into similarly invisible substances. For instance, we burn a pound of wood, and a few grains of ash remain, the rest has disappeared into air; as we say now, it has been converted into carbonic anhydride gas ; as Hooke said then, it has been dissolved by the air.

2. "That this action it (the air) performs not until the body be sufficiently heated."

In more modern phraseology, every combustible possesses its special igniting point—phosphorus 92° F., sulphur 482° F., and so on.

3. "That this action of dissolution produces or generates a very great heat, and that which we call *Fire.*"

4. "That this action is performed with so great a violence, and does so minutely act, and rapidly agitate the smallest parts of the combustible matter, that it produces in the diaphanous medium of the air the action, or pulse, of *Light*."

This would seem to indicate that Hooke considered light to be an intensified form of heat, and to be generated in the same manner, and to be a kind of very rapid motion.

5. "That the dissolution of sulphurous bodies is made by a substance inherent and mixed with the air, that is like, if not the very same with, that which is fixed in saltpetre."

Hooke had evidently traced the connection between certain actions produced by the air and by saltpetre or nitre; and he says it may be readily demonstrated that combustion is effected by that constituent of the air which is fixed in saltpetre. This is a remarkable assertion, because oxygen gas was not discovered until more than a century after the proposition of Hooke's theory; and we now know that nitre contains "fixed" in it the same substance—oxygen gas—which causes air to "dissolve" combustible bodies. It is probable that the connection between air and nitre may have been rendered the more probable in the minds of Hooke and his contemporaries by the knowledge that gunpowder will burn in a space devoid of air; thus, if sulphur and charcoal burn in air, and consume air in burning, and if nitre will cause them to burn out of contact with air, it would surely appear that nitre must contain air, or one of its components.

10. "That the dissolving parts of the air are but few, ... whereas saltpetre is a menstruum, when melted and red hot, that abounds more with these dissolvent particles, and therefore as a small quantity of it will dissolve a great sulphurous body, so wi l the dissolution be very quick and violent."

It was well known that if a piece of red-hot charcoal be thrown into melted nitre, it is consumed with great rapidity, while in the air it burns with far less readiness; hence Hooke infers that that particular component of air which causes it to support combustion exists in a condensed form in saltpetre. He also remarks, that if air be violently forced upon a piece of ignited charcoal by bellows it may be made to burn almost as rapidly as in melted nitre.

12. "It seems reasonable to think that there is no such thing as an *element of fire* . . . . but that that shining transient body called *flame* is nothing else but a mixture of air and the volatile sulphurous parts of dissoluble or combustible bodies."

Hooke asserts that this theory had been worked out by him several years earlier, and had been well supported by experimental means: he says, moreover, that he has here "only time to hint an hypothesis, which, if God permit me life and opportunity, I may elsewhere prosecute, improve, and publish." This he never did; but a young Oxford physician named John Mayow (b. 1645, d. 1679) eagerly accepted the theory, and adduced many experiments in support of it. Perhaps Mayow may have worked with Hooke during his residence in Oxford, and may have helped to adduce verifications of the then half-formed theory.

Mayow's experiments are contained in a treatise entitled " *Tractatus Quinque Medico-Physici quorum
primus agit de Sal-nitro et Spiritu Nitro-aëreo, Secun-
dus de Respiratione* .... Oxonii, 1674." The book is
altogether important, because the experiments which

FIG. 19 —JOHN MAYOW.
(From his "Tractatus Quinque Medico-Physici, 1674.")

it contains form the basis of pneumatic chemistry,
that is, the chemistry of gaseous bodies; it is also
distinguished by accurate reasoning and well-founded
generalisations. Had it been better known, it can
scarcely be doubted that the discovery of oxygen, and

of various gases made a century ago, would have been forestalled by many years.

Mayow calls the "dissolving parts" of the air and of nitre, which we now call oxygen gas, by the several names of *nitre-air*, *fire-air*, and *nitro-aërial spirit.* Air does not consist wholly of nitre-air, because when a candle is burnt in a closed vessel only a portion of the contained air is consumed. Nitre-air exists in large quantities in a condensed form in nitre ; hence combustible bodies mixed with nitre will burn under water, or in a vacuum. The acid of nitre contains all the nitre-air in nitre, but it does not inflame bodies so readily as nitre, because in it the nitre-air is surrounded by particles of water which tend to quench the burning body. Nitre-air is not combustible itself, neither does nitre contain any combustible substance, because it may be fused in a red-hot crucible, but no ignition will be observed to take place until a combustible body has been added. All acids contain nitre-air :—how curiously this contrasts with Lavoisier's name *oxygen*, from $o\xi v \varsigma \gamma \epsilon v v \alpha \omega$, which he gave to the gas, because he believed it to be an essential constituent of all acids. Sulphuric acid, according to Mayow, consists of nitre-air united with sulphur ; wines become sour and are changed into vinegar by the absorption of nitre-air from the atmosphere. It is the cause also of fermentation and putrefaction, and for this reason, substances when covered with fat or oil do not putrefy. During calcination metals increase in weight, and this increase is attributed by Mayow to absorption of nitric air ; thus calx of antimony is antimony *plus* nitre-air, and this is borne out by the fact that a substance absolutely similar to calx of antimony may be procured by treating the metal with the acid of nitre and evaporating. Again, rust of iron is iron united with nitre-air.

We come now to some of the first experiments in Pneumatic Chemistry. In one of his experiments Mayow supported a kind of ledge within a bell-jar full of air (see Fig. 20); upon the ledge he placed a piece of camphor, and fired it by concentrating the rays of the sun by a lens upon it. The camphor

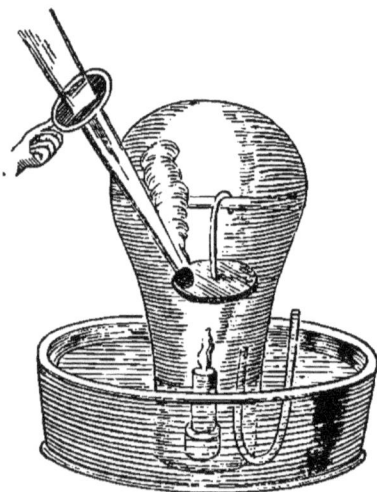

FIG. 20.—Early experiment in pneumatic chemistry.

ignited and burnt for some time; water then rose in the jar; and on again attempting to ignite the camphor he was unsuccessful. A lighted candle was also burned in the jar with the same result. Thus a part only of the air had been consumed, and the remainder was unable to support combustion. The siphon tube (shown on the right-hand side of the figure) was inserted at the commencement in order to render the height of the water the same inside and outside the tube, as stoppered air-jars were then unknown.

Thus it was clearly proved that air is diminished in bulk by combustion. In order to prove that respiration produces a similar result, Mayow tied a piece of moist bladder over the mouth of a jar (Fig. 21), and upon this he pressed a cupping-glass, so that the edges fitted air-tight. Within the cupping-glass he placed a mouse, and as the animal continued to breathe he noticed that the bladder was forced up more and more into the cupping-glass, proving that the air within it had been diminished by the respiration. Thus Mayow endeavoured to establish a connection between combustion and respiration. He also placed a mouse in a vessel standing over water, and noticed that the water rose in the jar as the respiration continued; and he found it impossible to ignite a combustible body in a jar of air in which a mouse had been suffocated. Again, he placed a mouse and a lighted candle together in a jar of air, and he noticed that the mouse only lived half as long as a mouse lived in the same bulk of air without the candle. Air deprived of its nitre-air was assumed to be lighter than nitre-air, because if a mouse is placed near the top of a closed vessel it dies sooner than if placed near the bottom.

FIG. 21.—Early experiment in physiological chemistry.

In 1672 Robert Boyle procured hydrogen gas by acting upon iron filings with an acid, and proved its inflammability; but he does not appear to have further studied its properties, and its discovery is always attributed to Cavendish, a century later. Boyle suggests that it probably consists of "the volatile sulphur of Mars, or of metalline steams participating in a sulphurous nature." Mayow also procured some of this gas by acting upon iron with dilute sulphuric acid, and he proves that it is not a supporter of life.

Mayow's second treatise is on Respiration, and he herein expresses' views far in advance of any of his predecessors. He proved that the nitre-air is alone concerned in respiration, and he asserts that this is absorbed by the blood, while the rest is rejected. It unites with combustible particles in the lungs, and thus produces animal heat. The lungs consist of a number of minute sack-shaped membranes through which the nitre-air passes to the blood.

We add the following *résumé* of Mayow's treatise, and of the position which it ought to occupy in the history of chemistry, from an article which we wrote on the subject a few years ago.

Mayow's work is remarkable in several respects. In it he conclusively proved that respiration and combustion are analogous processes; he upset the four-element theory by demonstrating the compound nature of air; and he recognized oxygen and nitrogen as clearly and almost as notably as they were recognized a hundred years later—the one the supporter of life and combustion, the principle of acidity, and the cause of fermentaticn and putrefaction, heavier

than atmospheric air; the other incapable of sup-
porting life or combustion, and lighter than atmospheric air. We find, moreover, in this work the
dawn of the idea of chemical affinity in the *fermentation*, which he speaks of as taking place between
nitre-air and combustible particles, and extending to
the production or destruction of things. Mayow even
employs some of the terms in general use in the
present day; thus he speaks of *affinitas*, existing
between acids, and earthy substances, and uses the
words *combinetur* and *combinentur* in speaking of the
congressus of different substances.

The treatise is characterised by much clear and
condensed thought, well-sustained argument, and
accurate reasoning; moreover, we seldom meet with
instances of too hasty generalisation, always the
dominant source of error in the early development
of a science. We further observe a great advance
towards that exact and discriminative mode of thought
which is necessary for the investigation of chemical
phenomena. The period in which Mayow wrote was,
as regards chemistry, a period of transition; there
was as yet no work on scientific chemistry, yet
Mayow's treatise approached more nearly to such
a work than that of any of his predecessors. The
works of previous writers in this direction belonged
to one of the three following classes : they were either
chemico-metallurgic, chemico-medical, or alchemical
treatises, or they partook of the nature of all three.
The publication of works on alchemy was fast waning
before the advances of the new philosophy; for as
superstition retreats, and as men begin to devote

their energies to the legitimate investigation of Nature, a false and chimerical art must of necessity cease to find votaries. Mayow was the first to discuss the intimate nature of an intangible body; other writers had treated of the air as a whole, but no one had endeavoured to ascertain the nature of its internal constitution, or to determine why it produces certain changes in surrounding bodies, upon what these changes depend, and the nature of the constituent or constituents of the air producing them. The old dogma of the elemental nature of the air was received as an absolute truth, although entirely unproven; it was thought that a theory which had been received since the earliest ages must of necessity be correct, and no attempt was made to disprove it.

We see from the above that it was the investigation of the nature of nitre which led to the knowledge of the constitution of the air, and to the first experiments in Pneumatic Chemistry. Mayow remarks at the commencement of his treatise, that so much had been written about nitre, that it would appear "*ut sal hoc admirabile non minus in philosophia, quam bello strepitus edcret; omniaque sonitu suo impleret;*" and when we remember its connection with the foregoing results we are almost inclined to agree with him.

CHAPTER XI.

The Theory of Phlogiston—Comparison with Hooke's Theory of Combustion—Early Ideas regarding Calcination—Stephen Hales—His Pneumatic Experiments—Boerhaave—Conclusion.

ABOUT the year 1669 we find the first dawnings of a theory which was proposed in order to connect together various chemical phenomena, and notably for the explanation of combustion, the common and most obvious of all chemical actions. This theory, known as the "Theory of Phlogiston," powerfully influenced chemistry for a century ; indeed, upon its ruins the structure of modern chemistry was raised by the labours of Lavoisier, Priestley, and Scheele. The proposers of this theory—John Joachim Beccher (b. 1625, d. 1682) and George Ernest Stahl (b. 1660, d. 1734) endeavoured to trace the cause of various phenomena of chemical change to the assimilation or rejection of what they called "*materia aut principium ignis non ipse ignis*"—not actual fire, but the principle of fire; a something not much unlike the pure elemental, celestial fire which a few Ancient and many Middle Age writers had feigned to exist. Stahl believed this *materia ignis* to be a very subtle, invisible substance, which neither burns nor glows ; its particles penetrate the most dense substances, and are agitated by a very rapid motion. When a body is burned it loses Phlogiston ; when a body is un-burned, if we may use such an expression, or de-oxidised, it assimilates Phlogiston ($\phi\lambda o\gamma\iota\sigma\tau\acute{o}\varsigma$, burnt). Thus if lead is heated for some length of time it is converted into a powdery substance which they called *calx of lead*, and we, *lead oxide;* the lead has lost Phlogiston, said Stahl. On the other hand, if this same calx of lead is heated with red-hot charcoal, it is de-oxidised and becomes lead again. It has now assimilated the Phlogiston, which it had before lost.

But here arose a difficulty. A metal was found to be heavier after calcination than before ; thus loss of Phlogiston led to gain of weight, which was altogether anomalous, and apparently incapable of explanation. But the Phlogistians were equal to the occasion : the supporters of a pet theory will create any number of the most vague and impossible hypotheses rather than yield up their darling to destruction : so, said they, Phlogiston is a principle of levity ; it confers negative weight, it makes bodies lighter, just as bladders attached to a swimmer lighten him.

The theory was applied as generally as possible :— thus sulphuric acid is produced by burning sulphur under certain conditions of oxidation ; the sulphur loses Phlogiston, and becomes heavier, like the metallic calx ; hence sulphuric acid is sulphur *minus* Phlogiston, while sulphur is consequently sulphuric acid *plus* Phlogiston. In fact, *loss of Phlogiston* was synonymous with what we call *oxidation ;* and *gain of Phlogiston* with *de-oxidation.* The existence of Phlogiston was so utterly unsupported by experimental proof that the theory could scarcely exist without many opponents. The endurance of the most false and chimerical theory is often really wonderful. The Phlogistians were attacked first in one direction, then in another, yet the theory continued to find supporters. Finally, as a last resource, hydrogen gas—recently investigated by Cavendish—was said to be Phlogiston; but this was so entirely different from the Phlogiston of Stahl that the theory was now seen on all sides to be fast giving way. At length Lavoisier, a century ago, conclusively disproved the theory by means which cannot be discussed here, because they belong to the more advanced history of the science.

How the crude, unscientific, illogical theory of Phlogiston could have arisen in the face of Hooke's admirable Theory of Combustion and Mayow's experiments in support of it, must always remain a mystery. It is probable that if Mayow had not died a young man, or if Hooke had found leisure to prosecute his views, the Theory of Phlogiston would never have been propounded. The theory has been much over-praised. The only service which it rendered to the science was that it introduced a certain amount of order and system, which was hitherto wanting. It led to the grouping together of certain classes of facts, and, to a slight extent, to the application of similar modes of reasoning to similar chemical phenomena. And although that reasoning

was altogether wrong, it seemed to indicate the means
by which, with a more perfect and advanced system,
chemistry might become an exact science subject to
definite modes of treatment.

We have more than once spoken of calcination,
which was indeed one of the most prominent operations
of old chemistry. Since the examination of the
process led to the proposal of just ideas concerning
the materiality of the air—most often denied by
Ancient and Middle Age writers—it may be well to
glance at the early ideas regarding calcination. Here
then was the dominant experiment in this direction :
I take a bright lustrous metal, tin or lead, melt it,
keep it in a molten state for a while, and it is converted
into powder which weighs more than the original
metal. Again I heat this same powder with charcoal,
and it becomes metal again ; yet nothing that can be
seen has been added to the metal, or taken away from
its calx. Geber defines calcination as "the pul-
verisation of a thing by fire, by depriving it of the
humidity which consolidates its parts." He observed
that the metal increases in weight during the operation, although "deprived of its humidity." Cardanus
asserted that the increase of weight in the case of
lead amounted to one-thirteenth the weight of the
metal calcined ; and he accounted for it on the supposition that all things possess a certain kind of life—
a *celestial heat,* which is destroyed during calcination ;
hence they become heavier, for the same reason that
animals are heavier after death, for the celestial heat
tends upwards. This idea was almost similar to that
of the Phlogistians, although published more than a

century before Beccher wrote his *Physica Subterranea.*
In 1629 Jean Rey, a physician of Bergerac, attempted
to discover the cause of increase, and attributed it to
the absorption of "thickened air" (*l'air espessi*) by the
metal during calcination. Lemery, as we have seen,
attributed the gain to the absorption of *corpuscules de
feu.* Afterwards came the *nitre air* of Mayow ; then
a century later the increase was proved to be due to
the union of the body with a constituent of the air
which Lavoisier named oxygen gas; and this gas
was first discovered by heating one of the calces (calx
of mercury), about which so much speculation had
been wasted, and so little experiment bestowed, by
earlier writers.

We are drawing towards the end of our subject,
but we think any account of the earlier history of
chemistry would be very incomplete without a notice
of the work of Dr. Stephen Hales (b. 1677, d. 1761).
In a number of Papers communicated to the Royal
Society, and afterwards published in a work entitled
Statical Essays, we find a variety of experiments by
Hales, chiefly relating to pneumatic chemistry. Herein
there is an account of "A specimen of an attempt
to Analyse the Air by a great variety of chymico-
statical experiments, which show in how great a proportion air is wrought into the composition of animal,
vegetable, and mineral substances, and withal how
readily it resumes its former elastic state when, in
the dissolution of those substances, it is disengaged
from them." In order to determine the quantity of
air disengaged from any substance during distillation
or fusion, Hales placed the substance in a retort, and

K

luted the retort to a large receiver with a small hole
at the bottom ; water was caused to occupy a known
space in the receiver, and the amount of air expelled
was estimated by noting the amount of water remaining in the receiver at the conclusion of the
experiment, after cooling.   Hales employed the appa-

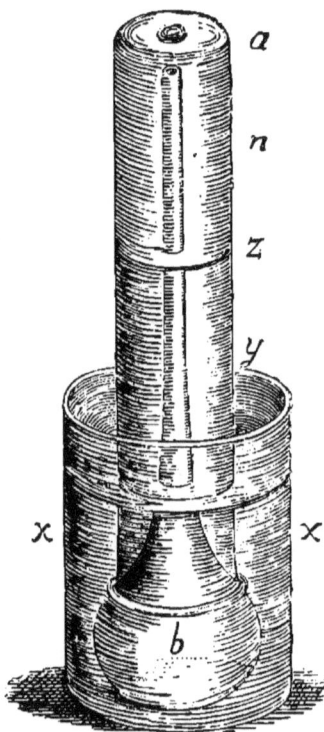

Fig. 22.—Hales' method of measuring
a gas.

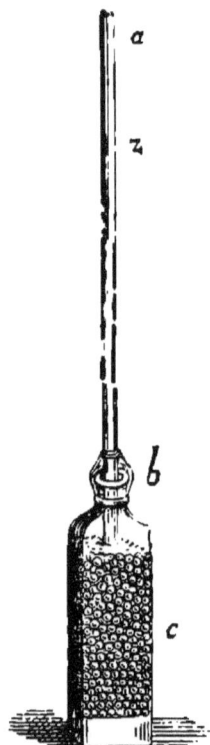

Fig. 23.—Measurement of the
elastic force of the gas produced
by fermenting peas.

ratus (Fig. 22) to measure the volume of· air generated by any kind of fermentation, also by the reaction
of one body upon another.

The substances undergoing fermentation were placed in *b*, and over the whole a vessel, *a y*, was inverted, closed below by water in the vessel, *x x*, and containing above a certain amount of air, to the level, *z*. If air were generated, the water in *a* sank (say to *y*) ; while if air were absorbed by the bodies in *b*, the water rose (say to *n*). Sometimes he placed different substances on pedestals in a jar of air, and ignited them, as Mayow had done, by a burning-glass, and noted the alteration in the bulk of air. He did this with phosphorus, brown paper dipped in nitre, sulphur, and other substances. If he required to act upon substances by means of a strong acid, he would place the substance in a suitable vessel on a pedestal in a known volume of air, standing over water, and would suspend over it a phial which could be emptied by pulling a string. These devices were closely copied by Priestley and Lavoisier in their experiments upon gaseous bodies. If a substance required to be heated violently, it was placed in a bent gun-barrel, *r r* (Fig. 24), one end of which was placed in a furnace, while the other was placed under a bell-jar, *a b*, full of water, inserted in the pail of water, *x x*. He distilled a number of substances, apparently taken at random, and determined the amount of gas evolved ; but he appears to have been at no pains to determine the nature of the gas, assuming it to be ordinary atmospheric air. Thus he distilled one cubic inch of lard, and collected thirty-three cubic inches of gas as the product of decomposition. Tallow, horn, sal-ammoniac, oyster-shells, peas, amber, camphire, and many other substances were similarly treated.

Two grains of phosphorus ignited in a closed vessel of air were found to absorb 28 cubic inches of air. 211 grains of nitre mixed with bone-ash yielded 90 cubic inches of gas; 54 cubic inches of water on

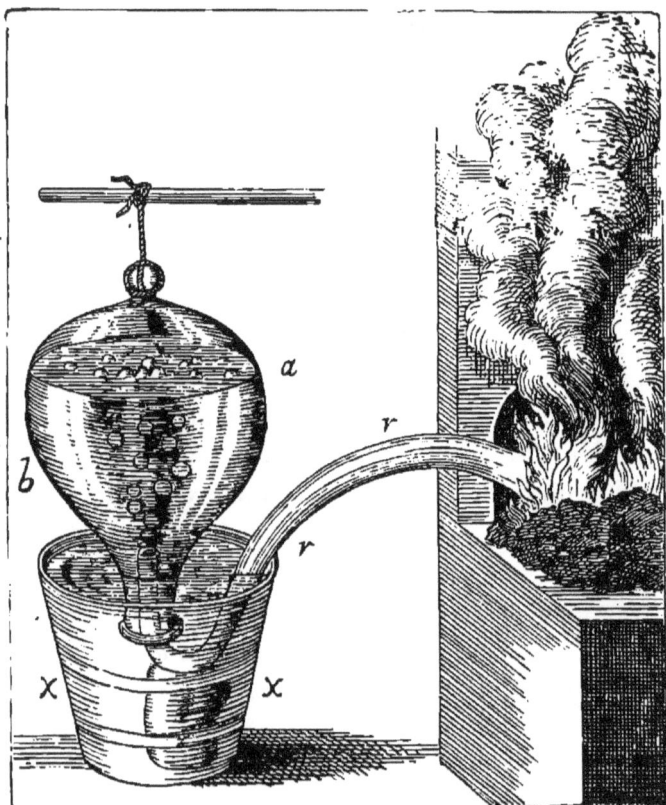

. FIG. 24.—Hales, pneumatic experiments.

boiling yielded 1 cubic inch of air. In order to measure the elastic force of the gas produced by fermenting peas, Hales filled a small, strong bottle, *c* (Fig. 23) with peas, filling up the interstices with water ; mercury to a depth of half an inch was then poured in, and of course remained at the bottom of the vessel, *c*. A long tube, *a z*, the lower end of which dipped beneath the mercury, was securely fastened into the mouth of the bottle, *b*, and fixed air-tight. In a few days' time the peas were in a state of fermentation, and the generated gas had forced the mercury to ascend in the tube *a z* to a height of 80 inches ; hence the gas in *c* was existing under a pressure of about 35 lbs. on the square · inch.

Hales also produced gases by various reactions. Thus he poured a cubic inch of sulphuric acid on half a cubic inch of iron filings : no effect took place until he had diluted the acid with water, when 43 cubic inches of *air* (as he calls it, in reality hydrogen gas) came off. Iron filings mixed with nitric acid, or with ammonia, or sulphur, were found to absorb air. A cubic inch of chalk treated with dilute sulphuric acid produced 31 cubic inches of *air* (in reality, carbonic anhydride gas). If space permitted, we could say much more of Hales' work. His experiments on respiration, and on various principles of vegetation, are exceedingly ingenious, and often accurate. It has often been said that Lavoisier created modern chemistry by the introduction of the balance into chemical experiments ; but here we find Hales weighing his substances, and measuring his gases, years before Lavoisier was born. Hales did not sufficiently investigate the nature of the various gases which he produced in the course of his experiments, but he

assuredly paved the way for many of the after-dis-
coveries of Priestley, Cavendish, and Lavoisier.

Dr. Hermann Boerhaave, of Leyden (b. 1668, d.
1738), was a contemporary of Hales. He was the
author of the first comprehensive system of chemistry :
a bulky quarto in two volumes, entitled *Elementa
Chemiæ*, which appeared in 1732, and which for many
years was the chemical text-book of Europe. In it
he defines chemistry as "an art which teaches the
manner of performing certain physical operations,
whereby bodies cognizable to the senses, or capable
of being rendered cognizable, and of being contained
in vessels, are so changed by means of proper instruments, as to produce certain determinate effects,
and at the same time discover the causes thereof, for
the service of various arts."

But, hold ! our task was to give some account of the
birth of chemistry, while a science with such a ponderous definition as the above is no longer infantile.
The babe has grown up about us until it has assumed
a tremendous individuality. The great discoveries of
the fathers of modern chemistry, Lavoisier, Scheele,
Priestley, Cavendish, Davy, need not be told here :
they belong to the later history of chemistry. We
have traced the science from its commencement in
the crude metallurgical, and other operations of the
ancients, to the time when a comprehensive system of
chemistry appeared. When we think of the vast
dimensions of the science of to-day, the numberless
text-books which are to be found in every language,
the great laboratories springing up in every country,

the immense amount of original research, we are carried back in spirit to those mistaken, but often grandly energetic, men who said to the disciples of their art :—

ORA!
LEGE, LEGE, LEGE, RELEGE. LABORA ;
ET INVENIES.

THE END.

LONDON : R CLAY, SONS AND TAYLOR PRINTERS.